ROWAN UNIVERSITY
CAMPBELL LIBRARY
201 MULLICA HILL RD.
GLASSBORO, NJ 08028-1701

D1608132

"The Earth Is Our Book"

RECENTIORES: LATER LATIN TEXTS AND CONTEXTS
James J. O'Donnell, Series Editor

Editorial Board
Paula Fredriksen, Boston University
James W. Halporn, Indiana University
E. Ann Matter, University of Pennsylvania
Carol Neel, The Colorado College
Stephen G. Nichols, The Johns Hopkins University
Mary Wack, Washington State University

Poetry and the Cult of the Martyrs: The Liber Peristephanon *of Prudentius*
    by Michael Roberts

Dante's Epistle to Cangrande
    by Robert Hollander

Macaronic Sermons: Bilingualism and Preaching in Late-Medieval England
    by Siegfried Wenzel

Writing Ravenna: The Liber Pontificalis *of Andreas Agnellus*
    by Joaquín Martínez Pizarro

Anacreon redivivus: *A Study of Anacreontic Translation in Mid-Sixteenth-Century France*
    by John O'Brien

The Whole Book: Cultural Perspectives on the Medieval Miscellany
    edited by Stephen G. Nichols and Siegfried Wenzel

Parody in the Middle Ages: The Latin Tradition
    by Martha Bayless

A Comedy Called Susenbrotus
    edited and translated by Connie McQuillen

The Poetry and Paintings of the First Bible of Charles the Bald
    by Paul Edward Dutton and Herbert L. Kessler

Rereading the Renaissance: Petrarch, Augustine, and the Language of Humanism
    by Carol Everhart Quillen

Lucian and the Latins: Humor and Humanism in the Early Renaissance
    by David Marsh

The Limits of Ancient Christianity: Essays on Late Antique Thought and Culture in Honor of R. A. Markus
    edited by William E. Klingshirn and Mark Vessey

Pierio Valeriano on the Ill Fortune of Learned Men: A Renaissance Humanist and His World
    by Julia Haig Gaisser

"The Earth Is Our Book": Geographical Knowledge in the Latin West ca. 400–1000
    by Natalia Lozovsky

The Correspondence of Johann Amerbach: Early Printing in Its Social Context
    selected, translated, edited, with commentary by Barbara C. Halporn

Magnus Felix Ennodius: A Gentleman of the Church
    by S. A. H. Kennell

"The Earth Is Our Book"

Geographical Knowledge in the Latin West ca. 400–1000

Natalia Lozovsky

Ann Arbor

THE UNIVERSITY OF MICHIGAN PRESS

G
89
L68
2000

Copyright © by the University of Michigan 2000
All rights reserved
Published in the United States of America by
The University of Michigan Press
Manufactured in the United States of America
♾ Printed on acid-free paper

2003   2002   2001   2000   4   3   2   1

No part of this publication may be reproduced, stored in a retrieval system, or transmitted in any form or by any means, electronic, mechanical, or otherwise, without the written permission of the publisher.

A CIP catalog record for this book is available from the British Library.

Library of Congress Cataloging-in-Publication Data

Lozovsky, Natalia.
    The earth is our book : geographical knowledge in the Latin West ca. 400–1000 / by Natalia Lozovsky.
       p.   cm. — (Recentiores)
    Includes bibliographical references (p.   ) and index.
    ISBN 0-472-11132-9 (cloth : acid-free paper)
    1. Geography, Medieval.   I. Title.   II. Series.
G89.L68   2000
910—dc21   99-50981

Acknowledgments

Over the years leading to the completion of this study, I have been fortunate to receive inspiration and support from many individuals and institutions. It gives me great pleasure to express my gratitude to them.

The ideas that formed the conceptual framework of this book originated during my postgraduate study at the Institute of History of Science and Technology in Moscow. I am grateful to my mentors Natalia Kuznetsova and Mikhail Rozov and to the members of the methodological seminar at the Institute of History of Science and Technology for many vigorous discussions that tested and improved my first, immature hypotheses.

At the University of Colorado at Boulder and at other places, I have received support and encouragement from many people who generously shared with me their vast knowledge of medieval history and culture. Boyd Hill, my dissertation adviser, with unfailing patience and good humor answered innumerable questions, read and corrected numerous drafts, and guided me in matters intellectual and practical from the beginning of my doctoral study to the completion of this book. Steven Epstein made perceptive and inspiring comments, always pointing to new ideas and opening new ways of inquiry. Ralph Hexter helped to improve this work by making many illuminating suggestions based on his deep understanding of medieval Latin and medieval culture.

Bruce Eastwood, always generous with help, advice, and encouragement, suggested that I look at Martianus commentaries, gave me the opportunity to use his microfilm collection, and contributed numerous valuable comments at every stage of this project. Rosamond McKitterick gave inspiration and help in many questions of early medieval paleography and intellectual history. The faculty and staff of Newnham College, Cambridge, provided a most congenial place to work. I am grateful to John

Contreni, who painstakingly read the entire manuscript and improved it by many perceptive comments. I also thank my anonymous readers for their helpful suggestions. For illuminating methodological discussions of early sciences, I thank Vera Dorofeeva-Lichtman and Alexei Volkov. Any errors and imperfections that still remain in this study despite the generous efforts of all these people are my own responsibility.

Various institutions provided funding for my research. Grants from the Department of History and Graduate School at the University of Colorado at Boulder and the Lowe Dissertation Fellowship, awarded by the School of Arts and Sciences at the University of Colorado at Boulder, enabled me to travel to England and France. A Bernadotte E. Schmitt Grant from the American Historical Association provided funds for microfilming. I am thankful to these institutions for their support.

I would also like to thank the interlibrary loan staff of the University of Colorado. They have been endlessly patient and supportive in answering my numerous requests. Thanks are also due to the staff of the University of Michigan Press, especially to the copyeditor, Jill Wilson, for the expertise and care with which they brought this book to publication.

Grateful acknowledgment is made to the Universiteitsbibliotheek in Leiden and the Biblioteca Medicea Laurenziana in Florence for permission to reproduce folios from MSS Leiden, Universiteitsbibliotheek, Voss. lat. F. 48, and Florence, Biblioteca Medicea Laurenziana, San Marco 190.

Finally, this study could not have been accomplished without the loving encouragement and help of my family. My debt to them can be neither measured nor expressed in words.

Contents

Abbreviations	ix
Introduction	1
1. Geographical Tradition in the Medieval System of Knowledge and Education	6
The Names	8
The Contexts	10
2. Exegesis and Geographical Tradition	35
How Can You See the Earth?	36
Paradise Revisited	50
Paradise and the Earth—Identifications	59
Paradise and the Earth—Associations	63
3. History and Geographical Tradition	68
Orosius	69
Jordanes	78
Bede	86
Richer	94
4. Studying and Teaching Geography	102
The Isidorean Tradition	103
Martianus' Geography at School	113
5. What Was Early Medieval Geography?	139
The Image of the Earth	139
How to Understand the Created World—General Ideas	141
How to Understand and Describe the Earth—"Geographers"	145
The Uses of Theoretical Geographical Knowledge	152

Conclusion	156
Selected Bibliography	161
General Index	179

Abbreviations

CCSL	Corpus Cristianorum, Series Latina
CSEL	Corpus Scriptorum Ecclesiasticorum Latinorum
MGH	Monumenta Germaniae Historica
AA	Auctores Antiquissimi
SRL	Scriptores Rerum Langobardicum et Italicarum
SRM	Scriptores Rerum Merovingicarum
PL	Patrologiae Cursus Completus, Series Latina, ed. J.-P. Migne

Introduction

Early medieval geographical tradition resembles Andersen's Ugly Duckling. It often gets scolded and cannot find a place of its own, all because people almost invariably misunderstand its true nature.

Historians of geography, looking for the origins of their discipline in the past, have long ago labeled early medieval learned tradition "geography in decline." According to them, the early Middle Ages display the lowest scientific level between two high points—the Greeks and the Renaissance.[1] Even those historians of geography who consider the medieval period at length, sometimes very sympathetically, share this low opinion about early medieval geography.[2]

The conception of the early Middle Ages as the period of cultural and

---

1. This is as true now as it was two centuries ago. See the following list, far from complete but, I believe, representative: R. de Vaugondy, *Essai sur l'histoire de la géographie* (Paris, 1755), 32–34; J. Löwenberg, *Geschichte der Geographie* (Berlin, 1840), 103–10; O. Peschel, *Geschichte der Erdkunde,* 2d ed. (Munich, 1877), 79–81; J. Keane, *The Evolution of Geography* (London, 1899), 34; S. Günther, *Geschichte der Erdkunde* (Leipzig and Vienna, 1904), 31–33; A. Hettner, *Die Geographie: Ihre Geschichte, ihr Wesen und ihre Methoden* (Breslau, 1927), 34–36; J. Schmithüsen, *Geschichte der geographischen Wissenschaft von den ersten Anfängen bis zum Ende des 18. Jahrhunderts* (Mannheim, Vienna, and Zurich, 1970), 46–48; A. Holt-Jensen, *Geography: Its History and Concepts,* trans. B. Fullerton (London, 1982), 11; G. J. Martin and P. E. James, *All Possible Worlds,* 3d ed. (New York, 1993), 53–56.

2. See J. Lelewel, *La géographie du Moyen Âge,* (Brussels, 1852), vol. 1, chaps. 1–3; C. R. Beazley, *The Dawn of Modern Geography: A History of Exploration and Geographical Science,* vol. 1, *From the Conversion of the Roman Empire to A.D. 900, with an Account of the Achievements and Writings of the Christian, Arab, and Chinese Travellers and Students* (London, 1897), 243; W. L. Bevan and H. W. Philott, *Mediaeval Geography: An Essay in Illustration of the Hereford Mappa Mundi* (London, 1873), x; J. K. Wright, *The Geographical Lore of the Time of the Crusades: A Study in the History of Medieval Science and Tradition in Western Europe* (New York, 1925), 44; G. H. T. Kimble, *Geography in the Middle Ages* (London, 1938), chap. 2; *Travel and Travellers in the Middle Ages,* ed. A. P. Newton (New York, 1962), chap. 2.

scientific decline has now become a thing of the past. Historians study early medieval culture in its own context and on its own terms, avoiding contemporary value judgments. However, early medieval geography has been largely left out of this methodological mainstream. Historians of medieval science seem to avoid it.[3] It rarely figures in discussions of medieval culture and mentality.[4] For literary historians and textual critics, early medieval geographical texts have philological rather than geographical value.[5]

Medieval cartography and medieval pilgrim travel accounts, as cultural phenomena, have enjoyed much more scholarly attention than the learned tradition represented in texts.[6] Even though some of the texts have been closely studied, early medieval geographical knowledge largely remains a no-man's-land, rarely visited, inhabited by very few scholars who wish to explore it for its own sake. The recent findings of these scholars, however, reveal the enormous potential interest of this field.[7] They also reveal that, in order for this potential to be realized, scholars must recognize the early

---

3. In two recent collections on Carolingian sciences, only one article in each treats geography: see W. Bergmann, "Dicuils *De mensura orbis terrae*", in *Science in Western and Eastern Civilization in Carolingian Times*, ed. P. L. Butzer and D. Lohrmann (Basel, Boston, and Berlin, 1993), 527–37, and U. Lindgren, "Geographie in der Zeit der Karolinger," in *Karl der Grosse und sein Nachwirken: 1200 Jahre Kultur und Wissenschaft in Europa*, vol. 1, *Wissen und Weltbild*, ed. P. Butzer, M. Kerner, and W. Oberschelp (Turnhout, 1997), 507–19. In a good recent treatment of early science, there is no separate chapter on medieval geography: see D. Lindberg, *The Beginnings of Western Science* (Chicago, 1992).

4. See J. Le Goff, *Medieval Civilization*, trans. J. Barrow (New York, 1988), 138–40; A. Gurevich, *Categories of Medieval Culture*, trans. G. Campbell (London and Boston, 1985); P. Zumthor, *La mésure du monde: Representation de l'espace au Moyen Age* (Paris, 1993).

5. See M. Manitius, *Geschichte der lateinischen Literatur des Mittelalters* (Munich, 1911), 1:675; M. L. W. Laistner, *Thought and Letters in Western Europe: A.D. 500 to 900* (London, 1931), 283–85; R. Quadri, introduction to Anonymus Leidensis, *De situ orbis libri duo*, ed. R. Quadri (Padua, 1974), xx n. 1; F. Brunhölzl, *Geschichte der lateinischen Literatur des Mittelalters*, vol. 1: *Von Cassiodor bis zum Ausklang der karolingischen Erneuerung* (Munich, 1975), 308.

6. On cartography, mainly after the year 1000, see numerous works by A.-D. von den Brincken, for instance, "Mappa mundi und Chronographia," *Deutsches Archiv für Erforschung des Mittelalters* 24 (1968): 118–86, and "'. . . ut describeretur universus orbis': Zur Universalkartographie des Mittelalters," *Miscellanea Mediaevalia* 7 (1970): 249–78; *The History of Cartography*, vol. 1, *Cartography in Prehistoric, Ancient, and Medieval Europe and the Mediterranean*, ed. J. B. Harley and D. Woodward (Chicago, 1987); E. Edson, *Mapping Time and Space: How Medieval Mapmakers Viewed Their World* (London, 1997). On pilgrim accounts see below, chap. 3.

7. For instance, previously unknown texts keep coming to light and add to our knowledge of early medieval geographical studies: see P. Gautier Dalché, "*Situs orbis terre vel regionum*: Un traité de géographie inédit du haut Moyen Age (Paris, B.N. latin 4841)," *Revue d'histoire des textes* 12–13 (1982–83): 149–79; O. Szerwiniack, "Un commentaire hiberno-latin des deux premiers livres d'Orose, *Histoire contre les pagans*," *Archivium latinitatis Medii Aevi (Bulletin Du Cange)* 51 (1992–93): 5–137. I thank Patrick Gautier Dalché, who drew my attention to the latter publication.

medieval geographical tradition for what it is, a unique phenomenon of medieval culture, a certain type of knowledge, very different from modern geography.[8]

It is fairly easy to make this first methodological step; however, this only helps us assert what medieval geographical tradition was *not*. How, then, should we define what we look for? How should we focus our search? An examination of a wide range of early medieval texts reveals that even though geographical knowledge lacked a definition or a disciplinary status and functioned in various contexts, it still possessed a certain self-sufficiency as a subject; special "geographical" tracts testify to this. Therefore, without trying to give a definition of geography in the early Middle Ages, we should follow the ideas implicit in geographical descriptions of the time, particularly in self-sufficient ones, such as geographical tracts or chapters in encyclopedias.

These texts usually include the description of the earth as a whole and of its various regions, which reflects the early medieval understanding of the subject conveyed by the word *geography,* a word derived from Greek and translated into Latin as *orbis descriptio.* For lack of a better term and to avoid the word *geography,* burdened as it is with modern associations, I shall call the medieval approach to this subject "geographical knowledge," or "geographical tradition," or "geographical studies." Far from being precise terms that would fix for us the definition and boundaries of this early medieval cultural phenomenon, these are rather tentative descriptions, fluid as the phenomenon itself.

Building on these assumptions, the following study considers the question, What kind of knowledge was early medieval geographical tradition?[9] How did medieval people perceive its goals and its place in their system of knowledge; what was its subject; how did they teach and study it?

8. This idea is recognized by medievalists and some historians of geography. See P. Gautier Dalché, "Tradition et renouvellement dans la représentation d'espace géographique au IXe siècle," *Studi medievali,* 3d ser., 24 (1983): 121–65, esp. 163–65, and "Un problème d'histoire culturelle: Perception et représentation de l'espace au Moyen Age," *Médiévales: Langues, textes, histoire* 18 (1990): 5–15, esp. 5–6. Robert Fuson suggests that geography took a thousand-year detour after Ptolemy, in R. H. Fuson, *A Geography of Geography: Origins and Development of the Discipline* (Dubuque, Iowa, 1969), 47. David Livingstone begins the story of geography with the great discoveries, in D. Livingstone, *The Geographical Tradition: Episodes in the History of a Contested Enterprise* (Oxford, 1992).

9. A similar question proved useful with regard to other medieval branches of knowledge that do not fit our ideas of science, such as medieval *computus.* See F. Wallis, "The Church, the World, and the Time: Prolegomena to a History of the Medieval *Computus,*" in *Normes et pouvoirs à la fin du Moyen Age,* ed. M.-C. Désprez-Masson (Montreal, 1990), 15–29, esp. 17; see also Wallis' admirable analysis of *computus* in "Ms Oxford St. John's College 17: A Mediaeval Manuscript in Its Context" (Ph.D. diss., University of Toronto, 1985).

Chapter 1 looks at how early medieval people named and classified this knowledge. Geography did not exist as an independent discipline; texts that we would call geographical functioned in various contexts—exegesis, history, the seven liberal arts—which shaped the character of particular geographical writings. Therefore, our search for geographical knowledge must begin by establishing its possible contexts.

Chapters 2 and 3 examine how two contexts, exegesis and history, shaped early medieval geographical tradition. Chapter 2 explores how the Bible and biblical exegesis set up certain paradigms that structured the perception and representation of the earth in geographical writings. Chapter 3 looks at geographical descriptions in the context of history writing—what role they played and how historians' goals shaped them.

Chapter 4, analyzing two influential traditions in studying and teaching geographical material at school, addresses the concerns of early medieval geographical education. Chapter 5 explores the main characteristics of theoretical geographical knowledge by putting them in the context of ontological and epistemological ideas of the time.

Early medieval geographical tradition is a vast and only slightly explored field. To approach its study, I had to limit my inquiry chronologically and thematically. Chronologically, the present work covers the period from ca. A.D. 400 to ca. A.D. 1000, beginning roughly at the time of Augustine, when new intellectual rules and norms were being established, and ending at the last century before the cultural and intellectual changes of the 1100s. Thematically, it concentrates on the theoretical geographical tradition of the early Middle Ages as represented in texts, acknowledging the limitations imposed by texts in terms of both physical transmission and content.

Maps, for many people almost synonymous with medieval geography, represent the most important omission that needs to be explained. Due either to the loss of many medieval books or to the specific ideas of medieval scholars, maps seem to constitute a separate group of sources. They often appear unaccompanied by texts; texts in which they do appear only partially coincide with those that form the subject of this study. For instance, none of the extant early medieval tracts specially devoted to geography contains a map. Thus early medieval maps will not be considered here as a separate group; they require detailed study and currently enjoy much more attention than texts.[10]

The medieval geographical knowledge represented by the learned tex-

---

10. Evelyn Edson's recent book devoted to medieval maps, *Mapping Time and Space*, considers many questions and reaches some of the conclusions relevant for the study of early medieval geographical tradition.

tual tradition was largely concerned with the timeless components of the *orbis descriptio,* such as continents, seas, and the like, and it more often than not ignored contemporary political units and divisions, essentially reproducing the Roman ones. Concentrating on theoretical rather than practical knowledge, this study virtually excludes many aspects related to practical geographical knowledge. The latter is a theme so vast and promising that it deserves a special treatment,[11] as do both the interrelation between practical and theoretical knowledge and the possible political implications of early medieval geographical tradition.

With respect to manuscript sources, this study could only include a small part of the early medieval geographical tradition—school commentaries to book VI of Martianus Capella. To understand the development of early medieval geographical knowledge, we need to study the entire early medieval manuscript transmission of geographical texts.[12] Apparently this enormous enterprise can here be only approached; its fulfillment has to remain a program for the future.

As our first approach, we need to consider the nature and connections of early medieval geographical tradition. Not only does this approach help us to find the proper niche for one particular ugly duckling, but by restoring geographical knowledge to its proper context, it contributes to our understanding of medieval intellectual culture.

---

11. A recent study has once more shown how significant the extent of Carolingian practical knowledge was: see C. R. Bowlus, *Franks, Moravians, and Magyars: The Struggle for the Middle Danube, 788–907* (Philadelphia, 1995).

12. As the example of other medieval "sciences" shows, the study of manuscript transmission enables scholars to change completely the picture of their character and development; see, for instance, the studies of early medieval astronomy by Bruce Eastwood and the studies of medieval *computus* by Faith Wallis: B. Eastwood, *Astronomy and Optics from Pliny to Decartes: Texts, Diagrams, and Conceptual Structures* (London, 1989) and "Invention and Reform in Latin Planetary Astronomy of the Eleventh Century," in *Publications of the Journal of Medieval Latin,* ed. M. Herren (Turnhout, forthcoming); F. Wallis, "Images of Order in the Medieval *Computus,*" in *Ideas of Order in the Middle Ages,* ed. W. Ginsberg (Binghamton, 1990), 45–68; and "Ms Oxford St. John's College 17." I thank Bruce Eastwood for sending me a copy of his forthcoming article.

CHAPTER 1

Geographical Tradition in the Medieval System of Knowledge and Education

The widespread assumption of historians of geography that their discipline already existed in the past has led them to scorn early medieval geographical tradition as the most imperfect predecessor of modern geography. Questioning this assumption, I propose to study the early medieval view of how geographical tradition fit into the system of knowledge of the time. The present chapter will begin the search for early medieval geographical knowledge by looking closely at how medieval scholars named and classified it. This will help us understand the actual contexts that shaped its development, and it will at the same time introduce the main texts and main directions of geographical studies in the early Middle Ages.

The order in which I shall address the contexts requires explanation; it is largely an abstraction, somewhat arbitrarily imposed on medieval material. Early medieval knowledge is hard to categorize, because for medieval scholars, all parts of it were tightly connected and could function in various ways and contexts. Nevertheless, the question of contexts in which geographical knowledge developed needs to be posed, even in this tentative fashion, for the contexts often defined how geographical tradition functioned within them.

I shall start by addressing the ideas of Augustine and Cassiodorus, who, by incorporating the secular knowledge of Antiquity into a Christian framework, justified geographical studies and connected them to several possible contexts. Then I shall turn to the study of several individual contexts, such as *physica* and the quadrivium. Finally, I shall address the tradition of special geographical tracts.

Before turning to the analysis of early medieval geographical studies, it is necessary to remind ourselves that this branch of learning, just like others in that period, built on the foundation created in antiquity and, more specifically, in the Roman world.[1] Roman writers of the first centuries A.D.—Pomponius Mela, Pliny, and Julius Solinus—as well as scholars of late antiquity, such as Macrobius and Martianus Capella (writing, respectively, in the late fourth–early fifth and the fifth century A.D.) provided the transmission of classical geographical ideas to the later centuries. Even without independent access to Greek writings, early medieval Latin scholars, reading and copying manuscripts of Pliny's *Natural History,* with its extensive geographical books, Pomponius Mela's *Chorography,* Julius Solinus' *Collection of Remarkable Things,* and others, still received much of the Greek legacy. This legacy included both theoretical questions (such as the spherical form of the earth, its size, its division into three continents) and detailed descriptions of regions.[2] As a legacy from Roman geographical writers, the scholars of late antiquity and the early Middle Ages received certain methods and principles of geographical description, such as the use of written authority. Whereas Pliny extensively used contemporary administrative information, combining it with material drawn from

---

1. A book about Roman geographical knowledge and particularly its connections to the early medieval tradition still remains to be written. A full treatment of this topic lies beyond the scope of the present study. For a traditional and in many respects outdated account of Roman geography see J. O. Thomson, *History of Ancient Geography* (New York, 1965). For a good, although brief, summary of the Romans' geographical contributions see C. Nicolet, *Space, Geography, and Politics in the Early Roman Empire* (Ann Arbor, 1991), chap. 3. Nicolet also provides a good fundamental bibliography. See also the sharp critique of the existing views of Roman geography and cartography in K. Brodersen, *Terra Cognita: Studien zur römischen Raumerfassung* (Hildesheim-Zurich-New York, 1995). The questions of classical legacy in medieval learning have been widely studied by many scholars. For extensive treatment, as well as bibliography, see, for instance, P. Riché, *Education and Culture in the Barbarian West, Sixth through Eighth Centuries,* trans. J. J. Contreni (Columbia, S.C., 1976).

2. Some Greek geographical works may have been available to late antique and early medieval scholars in translation. As we shall see later in this chapter, Cassiodorus recommended that his monks read Ptolemy, a Greek geographer of the second century A.D. However, that early medieval authors mention Greek geographers (or any other authorities) does not necessarily mean that they read them; they may have simply repeated from an intermediary source the name and information borrowed. To determine with any precision the real, rather than acknowledged, sources, one needs, ideally, to study the contents of the library of the individual in question or at least to establish, by analyzing the manuscript transmission, what texts he could have had access to. Such detailed study has only begun for early medieval geographical texts (see, for instance, numerous works by Patrick Gautier Dalché on various authors, quoted later in this study); as it continues, the picture of the sources and influences that emerges will in many ways be more realistic than that allowed by the traditional *Quellenforschung*. Since not all authors treated in this book have yet benefited from the former type of analysis, in some cases I had to focus primarily on the sources acknowledged by medieval scholars as an indication of their choice of a framework to follow.

earlier books, Mela and Solinus almost completely relied on earlier writers. Thus Mela and Solinus—and Pliny to a lesser degree—already used the methods for which early medieval scholars are still often criticized: they cut and arranged bits and pieces of material going back to older authors. In both content and method, knowledge represented in the works of these Latin geographers, as well as in the works of other Roman writers, formed the legacy that the Latin West would work with. As I shall show in this chapter, both the names for this type of knowledge and many of the contexts in which it was applied also represented the classical legacy. This heritage of classical geographical tradition would be recontextualized, partly rejected or changed, but largely preserved and transmitted by later scholars.[3]

The Names

Early medieval people did not have a definitive term that would designate the branch of knowledge that we now call geography. They seem to have rarely used the Greek-derived term *geographia*. Even though this word, rare in classical and late Latin (it is only attested in Cicero and Ammianus Marcellinus), did appear in early medieval manuscripts, before coming back in the fifteenth century,[4] both *geographia* and its derivatives, *geo-*

---

3. It would be a worthy project for the future to study, with a special focus on geography, the dissemination of these and other texts transmitting classical geographical ideas (such as Virgil's *Georgics* or Macrobius' commentary on Cicero's *Dream of Scipio*). These texts enjoyed a different degree of popularity in the early Middle Ages: for instance, there are only two or three pre-twelfth century manuscripts of Mela, at least sixteen of Pliny, around thirty of Solinus, and forty eight of Macrobius (some of the MSS containing maps.) On MSS transmission of these authors see B. Munk Olsen, *L'étude des auteurs classiques latins aux XIe et XIIe siècles,* vol. 2 (Paris, 1985). In addition, on the reception of Mela see C. M. Gormley, M. A. Rouse, and R. H. Rouse, "The Medieval Circulation of the *De Chorographia* of Pomponius Mela," *Mediaeval Studies* 46 (1984): 266–320; on medieval excerpts from *Natural History* that included geographical material see numerous works by Karl Rück, for instance, "Die Naturalis Historia des Plinius im Mittelalter. Exzerpte aus der Naturalis Historia auf den Bibliotheken zu Lucca, Paris und Leiden," *Sitzungsberichte der Königlich Bayerischen Akademie der Wissenschaften zu München, Philosophisch-historische Abteilung* 1 (1898): 203–318. On the reception of Solinus see M. E. Milham, "A Handlist of the Manuscripts of C. Julius Solinus," *Scriptorium* 37 (1983): 126–29. On Macrobius see B. Eastwood, "Manuscripts of Macrobius, *Commentarii in Somnium Scipionis*, before 1500," *Manuscripta* 38 (1994): 138–55, and W. H. Stahl, "Astronomy and Geography in Macrobius," *Transactions of the American Philological Association*, 73 (1942): 232–58.
4. Cicero, in the first century B.C., and Ammianus Marcellinus, in the fourth century A.D., use this word: see Cicero, *Letters to Atticus,* ed. and trans. D. R. Shackleton Bailey (Cambridge, 1965), 2.4.3; Ammianus Marcellinus, *Rerum gestarum libri qui supersunt,* vol. 1, ed.

graphicus (geographical) and *geographus* (geographer), required explanation. Ninth- and tenth-century Carolingian scholars, encountering these words in a geographical text by Martianus Capella, had to translate them from the Greek as "pertaining to the description of the earth."[5]

What did medieval people call the knowledge "pertaining to the description of the earth"? Cassiodorus, in his *Institutions,* uses the Greek term *cosmographia,* only attested in late Latin.[6] Certain manuscripts also use this word to describe the writings that we would call geographical. However, sometimes people felt this word also needed explanation, in which case they immediately provided the translation "the description of the world."[7] Another Greek term, *chorographia,* "the description of

---

V. Garthausen (1874; reprint, Stuttgart, 1968), 22.8.10. See also *Thesaurus linguae latinae,* vol. 6, pt. 2, col. 1907, and C. T. Lewis and C. Short, *A Latin Dictionary* (1879; reprint, Oxford, 1995), 811; the entry for *geographia* is absent in C. Du Cange, *Glossarium ad scriptores mediae et infimae latinitatis* (1687; rev. ed., Graz, 1954) and J. Niermeyer, *Mediae latinitatis lexicon minus* (Leiden, 1954). Du Cange (vol. 3, 58) only reports *geographare,* without a date, and Niermeyer (467) gives *geographus,* indicating its usage between A.D. 200 and 500. Cf. B. Guenée, *Histoire et culture historique dans l'Occident médiéval* (Paris, 1980), 166.

5. For *geographicus* see Martianus Capella, *De nuptiis Philologiae et Mercurii,* ed. A. Dick, corrected by J. Préaux (1925; reprint, Stuttgart, 1988), VI.609: *quo loco non puto transeundam opinationem Ptolemaei in geographico opere memoratam.* (Unless otherwise noted, translations of this text are from Martianus Capella, *The Marriage of Philology and Mercury,* trans. W. H. Stahl, R. Johnson, and E. L. Burge, in *Martianus Capella and the Seven Liberal Arts,* vol. 2 [New York, 1977].) Carolingian commentators usually explain the words *geographicus* and *geographia* by translating them from the Greek: see John Scottus, *Iohannis Scotti Annotationes in Marcianum,* ed. C. Lutz (Cambridge, 1939), 301.3, and Remigius of Auxerre, *Remigii Autissiodorensis Commentum in Martianum Capellam,* ed. C. Lutz (Leiden, 1962), 301.3: *IN GEOGRAPHICO OPERE id est in descriptione terrae.* For *geographus* see Niermeyer, *Mediae latinitatis lexicon minus,* 467: For *geographia* see, for instance, the manuscripts which, to my knowledge, no dictionary mentions: Paris, BN NAL 340, s. X, Cluny, fol. 77v and Paris, BN lat. 8674, s. X, Auxerre, fol. 71r, both at 301.3: *GEOGRAPHIA id est terrae scriptio.* (Here and in subsequent quotations I cite the glosses following the practice of Lutz, that is connecting the glosses to page and line numbers in Dick's edition.)

6. Cassiodorus, *Cassiodori Senatoris Institutiones,* ed. R. A. B. Mynors (Oxford, 1937), I.25. (Unless otherwise noted, translations of this text are from Cassiodorus, *An Introduction to Divine and Human Readings by Cassiodorus Senator,* trans. L. W. Jones [New York, 1969].) Cf. *Thesaurus linguae latinae,* vol. 4, col. 1083, and Lewis and Short, *Latin Dictionary,* 476; Du Cange has no entry for *cosmographia* in his *Glossarium.*

7. For instance, the work of Julius Honorius is called "cosmography": Verona, Bibl. Capit., II (2), s. VIex-VIIin, N. Italy, fol. 254r: *Explicit cosmografia iuli caesaris;* (see *Geographi latini minores,* ed. A. Riese [Heilbronn, 1878], XXXVII). For "cosmography" with an immediate explanation see Leiden, Universiteitsbibliotheek, Voss. lat. F 113, I, s. IX, fol. 1v: *Liber Ethico translato philosophico editus oraculo a Hieronimo presbytero delatum ex cosmographia, id est mundi scriptura* (underlined by me). The catalogue of the library of Bobbio, s. X, mentions a book on cosmography: *librum I cosmographia* (see G. Becker, *Catalogi bibliothecarum antiqui* [Bonn, 1885], 32, 471).

places," which appears in a ninth-century manuscript of Pomponius Mela, the Roman geographer of the first century A.D., does not seem ever to be applied to the whole field of knowledge.[8] The same seems to be true for Latin expressions that define the topics of writings describing the earth: "on the location of the earth," "the location of the entire earth and its various regions," or "the measurement of the earth."[9]

The fact that medieval scholars used so many terms that often needed explanation reflects the uncertainty of the field itself—its lack of definition and the fluidity of its boundaries as a discipline. No early medieval classification of sciences mentions knowledge about the earth as an independent field. If it figures at all in the discussions of the uses and hierarchy of knowledge, it is always in the context of another subject. Only by looking at these contexts can we establish how geographical tradition fit into the early medieval system of knowledge.

The Contexts

Augustine and the Justification of Geographical Studies

The system of knowledge that incorporated classical secular studies into the new Christian context was created in late antiquity. Augustine (354–430) launched its development, when he recognized, in his *On Christian Doctrine,* that certain secular studies were necessary for understanding the Bible, the source of divine wisdom and the goal of true knowledge. According to Augustine, to follow the sometimes obscure figurative language of the Bible, one should not be ignorant of "the natures of things" mentioned there. One should know certain things about the created world, such as "lands and animals, or grasses, trees and stones, or unknown metals, or any other specific things" mentioned in the Scripture, because otherwise one would not fully understand the sacred text.[10] A Christian

---

8. Vatican, Vat. lat. 4929, s. IX, fol. 149v: *Pomponii Melae de chorographia;* fol. 188r: *Pomponii Melae de chorographia libri tres expliciti feliciter.* Cf. Lewis and Short, *Latin Dictionary,* 325, citing Lactantius as the latest author.

9. *De situ orbis libellum* (in Leiden, Voss. lat. F 113, II—see Anonymus Leidensis *De situ orbis* Proemium 1); *totius orbis diversarumque regionis situs* (in Paris, BN lat. 4841, s. IX, Septimania—see Gautier Dalché, "Situs orbis terre", 162); *de mensura orbis terrae* (in Paris, BN lat. 4806, s. IX, Reims—see *Dicuili liber de mensura orbis terrae,* ed. J. J. Tierney and L. Bieler [Dublin, 1967]; unless noted otherwise, in subsequent quotations of Dicuil's text page numbers refer to the parallel translation given in this edition).

10. Augustine, *De doctrina christiana,* ed. J. Martin, CCSL, 32 pp. 1–167 (Turnhout, 1962), II.24 (trans. mine): *Rerum autem ignorantia facit obscuras figuratas locutiones, cum igno-*

should know at least as much as a non-Christian "about the earth, the heavens, and the other elements of this world," so that when he preached about the higher truths revealed in the Bible, he would not compromise them by ignorance of these mundane matters.[11]

Thus, Augustine regarded geographical information as part of knowledge about nature, or the created world, which fell into the category of *scientia*, "the knowledge of human things." He also indicated that this secular knowledge had to serve the understanding and preaching of the biblical truth, that is, essentially, biblical exegesis. Geographical knowledge, forming part of *scientia*, was meant to support *sapientia*, "the knowledge of divine things."[12]

Relating geography to knowledge about nature, Augustine also pointed out its connections to history. He justified historical studies in the same way as geographical studies, by their necessity for understanding the Scripture.[13] He often put them together; for instance, he compared the achievements of historical writers like Eusebius, who he claimed "has written a history because of the questions in the divine books that demand its use," and the task of those who explained the geographical locations, animals, and other natural phenomena mentioned in Scripture.[14]

Augustine also stressed the similarities in the methods that the two sub-

---

ramus uel animantium uel lapidum uel herbarum naturas aliarumque rerum.... See also II.59: *... ut quoscumque terrarum locos quaeue animalia uel herbas atque arbores siue lapides uel metalla incognita speciesque quaslibet scriptura commemorat, ea generatim digerens, sola exposita litteris mandet.* Unless otherwise noted, translations of this text are from Augustine, *On Christian Doctrine*, trans. D. W. Robertson (New York, 1958).

11. Augustine, *De Genesi ad litteram*, ed. J. Zycha, CSEL 28–1 (Vienna, 1894), I.19, p. 28: *Plerumque enim accidit, ut aliquid de terra, de caelo, de ceteris mundi huius elementis ... non christianus ita noverit, ut certissima ratione uel experientia teneat. turpe est autem nimis et perniciosum ac maxime cauendum, ut christianum de his rebus quasi secundum christianas literas loquentem ita delirare audiat....* Unless otherwise noted, translations of this text are from Augustine, *The Literal Meaning of Genesis*, trans. J. H. Taylor, 2 vols. (New York, 1982). Page numbers here and in subsequent citations of Augustine *De Gen. ad lit.* refer to Zycha's edition.

12. On human and divine knowledge, the categories that would become traditional in the Middle Ages, see Augustine, *De Trinitate*, ed. W. J. Mountain and F. Glorie, CCSL 50A (Turnhout, 1968), XIV.1: *... rerum diuinarum scientia sapientia proprie nuncupetur, humanarum autem proprie scientiae nomen obtineat ...*

13. Augustine *De doctrina christiana* II.42: *Quicquid igitur de ordine temporum transactorum indicat ea quae appellatur historia, plurimum nos adiuuat ad libros sanctos intellegendos, etiamsi praeter ecclesiam puerili eruditione discatur.*

14. Ibid., II.59 (trans. p. 74): *... et quod Eusebius fecit de temporum historia propter diuinorum librorum quaestiones, quae usum eius flagitant; ... sic uideo posse fieri, si quem eorum, qui possunt, ... ut quoscumque terrarum locos quaeue animalia uel herbas atque arbores siue lapides uel metalla incognita speciesque quaslibet scriptura commemorat....*

jects used and in the things they described. Geography, among other studies of nature, uses the method of demonstration and discusses the present. History uses description and discusses the past. However, description of the past is close to geographical demonstration, that is, the description of places.[15]

Both the study of time and the study of space, according to Augustine, address things instituted not by men but by God: "Thus he who narrates the order of time does not compose it himself; and he who shows the location of places or the natures of animals, plants, or minerals does not show things instituted by men; and he who demonstrates the stars and their motion does not demonstrate anything instituted by himself or other men. . . ."[16] People, while they observe all these things instituted by God and arrive at their conclusions by means of logical reasoning and the "discipline of disputation," should keep in mind that the ultimate truth of conclusions is instituted by God and only observed by humans, "so that they might learn or teach it."[17]

Emphasizing the relations between historical and geographical studies and their necessity for understanding Scripture, Augustine put into the Christian framework what had been a traditional connection since antiquity. In practice, late classical and early medieval historians included geographical introductions and digressions in their writings, following a tradition well-established in historical literature since Herodotus and familiar to the Latin West from the works of Caesar, Livy, and Tacitus, to name only a few. In theory, historians could rely on several conceptual frameworks that justified and recommended the connections between geographical and historical matters.

On the philosophical level, geographical and historical knowledge had to be related because the matters they described, place and time, were related. They came into existence together, created by God when the world

---

15. Ibid., II.45: *Est etiam narratio demonstrationi similis, qua non praeterita, sed praesentia indicantur ignaris. In quo genere sunt quaecumque de locorum situ naturisque animalium, lignorum, herbarum, lapidum aliorumque corporum scripta sunt.* On astronomy, Augustine writes (ibid., II.46), *Siderum autem cognoscendorum non narratio, sed demonstratio est.* . . .

16. Ibid., II.50 (trans. p. 68, modified by me): *Sicut enim qui narrat ordinem temporum, non eum ipse componit, et locorum situs aut naturas animalium uel stirpium uel lapidum qui ostendit, non res ostendit ab hominibus institutas, et ille qui demonstrat sidera eorumque motus, non a se uel ab homine aliquo rem institutam demonstrat.* . . .

17. Ibid., II.48 (trans. mine): *disputationis disciplina;* II.50 (trans. p. 68): *Ipsa tamen ueritas conexionum non instituta, sed animaduersa est ab hominibus et notata, ut eam possint uel discere uel docere; nam est in rerum ratione perpetua et diuinitus instituta.*

was made.[18] Early medieval thought perceived place and time as two of the most fundamental terms describing the world and its existence. Discussing the beginnings of the world, medieval scholars borrowed from classical philosophy the idea of ten basic categories: *usia* (substance), quantity, qualification, relation, place, time, being-in-a-position, having, doing, and being-affected.[19] Putting this question in the context of logic and debating about the categories' existence and their applicability to God, all medieval scholars seem to have agreed that the ten categories, place and time among them, could be used to describe the created world.[20]

This philosophical and logical tradition may have contributed to the approach of early medieval rhetoric. Just as philosophers described the world in terms of place and time, rhetoricians did the same for works of literature. Following the lessons of classical rhetoric, medieval scholars began their discussions with an introduction *(accessus ad auctores)* that gave the reader initial information about the author and his work.[21] Among other matters, the *accessus* could treat details about the work called "the seven circumstances" *(circumstantiae rerum),* which necessarily included both place and time of the work's composition.[22]

Thus connecting place and time as basic categories or definitions,[23] both philosophy and rhetoric contributed to the understanding of the sub-

---

18. For the discussion of the creation of space and time see Augustine, *De civitate Dei,* ed. B. Dombart and A. Kalb, 4th ed. (Teubner, 1878–79; reprint, CCSL 47–48, Turnhout, 1955), XI.6–8. Translations of this text are from Augustine, *Concerning the City of God against the Pagans,* trans. H. Bettenson (New York, 1972).

19. I use the translations of Aristotle's terms suggested by J. Marenbon in *From the Circle of Alcuin to the School of Auxerre* (Cambridge, 1981), 12. For a summary of versions of Aristotle's *Categories,* available to scholars in the early Middle Ages, see Aristotle, *Categoriae vel Praedicamenta,* ed. L. Minio-Paluello. Aristoteles latinus, I, 1–5 (Bruges and Paris, 1961), x–xxiii and lxvii–lxxx, and Marenbon, *From the Circle,* 16–20.

20. On different opinions see Marenbon, *From the Circle,* 24–29 and 50 on Boethius and Augustine, 72–73 and 75–77 on John Scottus. On place and time in John Scottus see M. Cristiani, "Le problème du lieu et du temps dans le livre Ier du 'Periphyseon,'" in *The Mind of Eriugena: Papers of a Colloquium, Dublin, 14–18 July 1970,* ed. J. J. O'Meara and L. Bieler (Dublin, 1973), 41–47.

21. On the *accessus* see E. A. Quain, *The Medieval Accessus ad Auctores* (New York, 1986).

22. Fortunatianus *Ars rhetoricae* II.1: *reperto statu quid consideramus? totam materiam per septem circumstantias; cur non statim dividimus? quoniam prius universam causam confuse considerare debemus, tunc omnia, quae reperta sunt, capitulatim quaestionibus ordinare; quae sunt circumstantiae? persona, res, causa, tempus, locus, modus, materia.* This passage is cited in H. Backes, *Die Hochzeit Merkurs und der Philologie: Studien zu Notkers Martian-Übersetzung* (Sigmaringen, 1982), 32. See ibid. on the seven *circumstantiae* in classical and medieval rhetoric.

23. On categories as definitions or *circumstantiae* see Cristiani, "Le problème," 42–43.

jects that treated place and time, geography and history, as interrelated. Classical rhetorical theory, which served as the model for medieval theory, advised historical and geographical digressions, usually mentioning them together. Cicero felt that both geographical and historical details helped the explanation: "The account of things requires chronological order and the description of places."[24] Priscian also advised the orator to describe times and places.[25] Supported by these frameworks, the connections between geographical and historical knowledge, traditional since antiquity, continued in the early Middle Ages. Historical writings continued to provide a context for geographical descriptions. Medieval maps represented both time and space and often occurred in manuscripts dealing with calendrical computations and history.[26]

Augustine, in his justification of geography (among other secular studies), put these traditional relations into a Christian framework. He also secured the place of geography in the Christian system of knowledge. Suggesting knowledge about nature as a most appropriate context for geographical studies, he emphasized the connections between geography and history. Both played an important role in Christian knowledge, preparing the ground for understanding the Bible. Augustine's justification of secular studies influenced the way the medieval system of knowledge was formed and taught at schools, including geographical material.

Cassiodorus and Geographical Studies at Vivarium

The plan of Christian studies proposed by Augustine was developed by Cassiodorus (ca. 490–580) in his *Institutions*. This work, written as an instructional program for the monks at Vivarium, can give us an idea of how geography fit into his system of knowledge and education. Cassiodorus, along with Augustine, Boethius (480–524), and others, established the classification of knowledge widely accepted in the early Middle

---

24. Cicero *De oratore* II.15.63: *Rerum ratio ordinem temporum desiderat, regionum descriptionem.* This passage is cited in H.-J. Witzel, *Der geographische Exkurs in den lateinischen Geschichtsquellen des Mittelalters* (Frankfurt, 1952), 31. For a detailed treatment of the place of geographical description in classical and medieval rhetoric see ibid., 22–48.

25. Priscian *Praeexercitamina* 10 (cited in Witzel, *Der geographische Exkurs,* 32): *oratio colligens et praesentans oculis quod demonstrat. Fiunt autem descriptiones tam personarum quam rerum et temporum et status et locorum et multorum aliorum; . . . locorum, ut litoris, campi, montium, urbium.*

26. For a detailed study of maps in the context of *computus* manuscripts, as well as in the context of manuscripts containing historical works, see Edson, *Mapping Time and Space.* See also the pioneering studies by Anna-Dorothee von den Brincken, the most important of which are listed by Edson (194).

Ages. Essentially drawn from Boethius and ultimately going back to the combination of Platonic and Aristotelian views, this classification defined philosophy—that is, the sum of all knowledge—as follows: "Philosophy is the probable knowledge of divine and human things insofar as it may be attained by man. According to another definition, philosophy is the art of arts and discipline of disciplines."[27] One of the two major divisions of philosophy—speculative philosophy—was further divided into natural philosophy, which discussed the natures of material things; theoretical philosophy, which considered abstract quantity; and divine philosophy, which treated the nature of God.[28] Theoretical philosophy, in its turn, consisted of the four disciplines of the quadrivium (arithmetic, music, geometry, and astronomy).

Even though none of these divisions explicitly mentioned geographical questions, Cassiodorus included them in his educational program. The list of readings that Cassiodorus suggested for his monks named texts most likely available at the Vivarium library and possibly used to teach geography.[29]

According to his definition of philosophy, Cassiodorus divided the *Institutions* into two books—the first devoted to divine letters, the second to secular. Geographical matters figure in both books in various contexts.

In book I, Cassiodorus discusses the Bible as the source of all wisdom and the disciplines that help understand it, such as exegesis and hagiography. History, which also falls into this useful category, is the first context that may have included geographical material at Vivarium. Among the writers Cassiodorus recommends in this section is Orosius, who wrote a long geographical introduction to his *Histories against the Pagans* (416–17).[30] However, Cassiodorus says nothing about Orosius' treatment of geography.

---

27. Cassiodorus *Institutiones* II.3.5 (trans. p. 160): *Philosophia est divinarum humanarumque rerum, in quantum homini possibile est, probabilis scientia. aliter, philosophia est ars artium et disciplina disciplinarum.*

28. Ibid., II.3.6.

29. On Cassiodorus' library see M. Cappuyns, "Cassiodore," in *Dictionnaire d'histoire et de géographie ecclésiastique,* vol. 11 (Paris, 1949), cols. 1349–1408, esp. cols. 1389–91; P. Courcelle, *Late Latin Writers and Their Greek Sources,* trans. H. E. Wedeck (Cambridge, Mass., 1969), 376–409, and "Histoire d'un brouillon cassiodorien," *Revue des etudes anciennes* 44 (1942): 65–86; J. J. O'Donnell, *Cassiodorus* (Berkeley, Los Angeles, and London, 1979), 207–21.

30. Orosius, *Historiarum aduersum paganos libri VII,* ed. K. Zangemeister, CSEL 5 (Vienna 1882; reprint, Hildesheim, 1967), *editio maior,* I.2. Unless otherwise noted, translations of this text are from Orosius, *The Seven Books of History against the Pagans,* trans. R. J. Deferrari (Washington, 1964).

Only once in his treatment of history does Cassiodorus refer to geographical questions—when, recommending Marcellinus, he mentions that this author devoted some attention to the description of places: "Marcellinus . . . has traversed his journey's path in laudable fashion, completing four books on the nature of times and the location of places with most decorous propriety. . . ."[31] Cassiodorus possibly means Marcellinus Comes (fl. ca. 534), and the work in question does not seem to have survived.[32]

Without specifically saying why one should know "the location of places," Cassiodorus connects it to "the nature of times," thus implying that geographical material may be somehow useful for the understanding of history. And Cassiodorus reserved a special role for history in the instruction of his monks: all historical writings that he recommended had to instruct them in heavenly affairs by assigning the causes of events and changes to the will of God.[33]

Next in book I, Cassiodorus mentions geographical questions in the context of cosmography, justifying these studies in the same way as Augustine did. The monks, he says, should have "some notion of cosmography, in order that you may clearly know in what part of the world the individual places about which you read in the sacred books are located."[34] However, the readings he recommends include much more than simply the locations mentioned in the Bible.

The first author Cassiodorus names is Julius Orator, who "has treated the seas, the islands, the famous mountains, the provinces, the cities, the rivers, the peoples, with a fourfold division in such a way that his book lacks practically nothing which is recognized as pertaining to a knowledge of cosmography."[35] This Julius Orator is also known as Julius Honorius,

---

31. Cassiodorus *Institutiones* I.17.1 (trans. p. 115, modified by me): *Marcellinus etiam, quattuor libros de temporum qualitatibus et positionibus locorum pulcherrima proprietate conficiens, itineris sui tramitem laudabiliter percurrit.* . . .

32. Later, at *Institutiones* I.17.2, Cassiodorus mentions another work by Marcellinus, the *Chronicon*, which covered the period from Theodosius to Justinian, and which has survived. Courcelle (*Late Latin Writers*, 374 n. 69) suggests that MS Oxford, Bodleian, Auct. T. 2. 6, s. VI, containing Marcellinus' work and Eusebius' *Chronica* in Jerome's translation, may have come from Vivarium.

33. Cassiodorus *Institutiones* I.17.1.

34. Ibid., I.25.1 (trans. p. 125): *Cosmographiae quoque notitiam vobis percurrendam esse non immerito suademus, ut loca singula, quae in libris sanctis legitis, in qua parte mundi sint posita evidenter cognoscere debeatis.*

35. Ibid. (trans. modified by me): *quod vobis proveniet absolute, si libellum Iulii Oratoris, quem vobis reliqui, studiose legere festinetis; qui maria, insulas, montes famosos, provincias, civitates, flumina, gentes ita quadrifaria distinctione complexus est, ut paene nihil libro ipsi desit, quod ad cosmographiae notitiam cognoscitur pertinere.*

the author of *Cosmography*, who might have lived in the fourth or fifth century A.D.[36]

This work, which Julius Honorius, "the skilled and undoubtedly the most learned teacher," dictated to his disciple, was accompanied by a map and represented a list of names—of seas, islands, provinces, cities, and peoples.[37] It resulted from and was intended for teaching. It may have been used as a manual not only at Vivarium but also in other sixth- and seventh-century Italian centers of learning.[38]

The next "cosmographical" writer Cassiodorus recommends in this section is Marcellinus, an author of historical works who, he says, "should be read with equal care." Cassiodorus notes that Marcellinus "has described the city of Constantinople and the city of Jerusalem in four short books in considerable detail."[39] Thus Cassiodorus includes into his notion of cosmography the descriptions of cities as well as lands or seas.

Along with texts, Cassiodorus mentions a map—"the concisely composed *Picture of the World* of Dionysius"—which he recommends to the monks "in order that you may contemplate almost as an eyewitness the things which you perceived with your ears in the above-mentioned book."[40] This map, which probably accompanied the description of the earth composed by Dionysius Periegetes, a second-century Greek writer, has not survived in the known manuscripts.[41]

---

36. The oldest manuscript of *Cosmography* (Paris, BN lat. 4808, fols. 53r–63r, s. VI, Italy) gives two names—Julius Honorius and Julius Orator. (See C. Nicolet and P. Gautier Dalché, "Les 'quatre sages' de Jules César et la 'mesure du monde' selon Julius Honorius: Réalité antique et tradition médiévale," *Journal des savants* [1986]: 157–218). On Julius Honorius' date and the manuscripts see *Geographi latini minores,* xxi, and xxxvi–xxxix and Nicolet and Gautier Dalché, "Les 'quatre sages,'" 162, and 184–88.

37. Julius Honorius *Cosmographia,* in *Geographi latini minores,* ed. A. Riese, 24–55, 50–51: *Et ut haec ratio ad conpendia ista deducta in nullum errorem cadat, sicut a magistro dictum est, hic liber excerptorum ab sphaera ne separetur. . . . Haec omnia in describtione recta orthographiae transtulit publicae rei consulens Iulius Honorius magister peritus atque sine aliqua dubitatione doctissimus. . . .*

38. Gautier Dalché argues that this could be the case for the seventh century but not the sixth. He cites two extant manuscripts produced in Italy—Verona, Bibl. Capit. II (2), s. VIex–VIIin, and Paris, BN lat. 4808, s. VI— and two exemplars for later collections (Nicolet and Gautier Dalché, "Les 'quatre sages,'" 187–88).

39. Cassiodorus *Institutiones* I.25.1 (trans. p. 125): *Marcellinus quoque, de quo iam dixi, pari cura legendus est; qui Constantinopolitanam civitatem et urbem Hierosolimorum quattuor libellis minutissima ratione descripsit.* This book has not survived.

40. Ibid., I.25.2 (trans. p. 125, modified by me): *Deinde Penacem Dionisii discite breviter comprehensum, ut quod auribus in supradicto libro percipitis, paene oculis intuentibus videre possitis.*

41. Beazley, *Dawn,* 1:390; Courcelle, *Late Latin Writers,* 395; P. Gautier Dalché, "De la glose à la contemplation: Place et fonction de la carte dans les manuscrits du haut Moyen Age," in *Testo e immagine nell'alto medioevo* (Spoleto, 1994), 2:693–764, esp. p. 695. The

The last cosmographer Cassiodorus recommends is Ptolemy, the second-century Greek geographer and astronomer. The library of Vivarium must have possessed his *Geography* in Latin translation.[42] It is interesting that Cassiodorus emphasizes the description of places as the most useful feature of Ptolemy's book and completely ignores its mathematical component, the measurements: "Then, if a noble concern for knowledge has set you on fire, you have the work of Ptolemy, who has described all places so clearly that you judge him to have been practically a resident of all regions, and as a result you, who are located in one spot, as is seemly for monks, traverse in your minds that which the travel of others has assembled with very great labour."[43]

In book II, devoted to secular letters, Cassiodorus treats the seven liberal arts. Of all geographical questions, he there only mentions the measurement of the earth, in the chapter on geometry—to explain the name of this discipline and its practical applications. As authorities in this question, he refers to Roman scholars Marcus Terentius Varro and Censorinus.

Varro, the Roman encyclopedic writer of the first century B.C. and, according to Cassiodorus, "the greatest expert of Latins," had connected the origin of geometry's name with land surveying and "the measuring of the entire earth."[44] The book Cassiodorus recommends is most likely the *Disciplines,* no longer extant. Censorinus "has carefully described the size

---

identification of Dionysius as the author of *Periegesis* rather than Dionysius Exiguus, a Scythian monk, whom Cassiodorus had earlier praised for his learning, seems more reasonable (for Exiguus see Jones' note in Cassiodorus, *An Introduction,* 125 n. 4). On the role of maps in medieval education see P. Gautier Dalché, *La "Descriptio mappe mundi" de Hugues de Saint-Victor: Texte inédit avec introduction et commentaire* (Paris, 1988), 95–99, and "De la glose," 696–97; and M. Kupfer, "Medieval World Maps: Embedded Images, Interpretive Frames," *Word and Image* 10 (1994): 262–88, esp. 264–76.

42. Although we have no direct evidence for this, Boethius himself may have translated Ptolemy's *Geography* as well as his *Astronomy* (we know about the latter from Cassiodorus *Variae* I.45.4: *Translationibus enim tuis Pythagoras musicus, Ptolemaeus astronomus leguntur Itali.*) Vivarium's own translators may also have performed the task: see Laistner, *Thought and Letters,* 100; W. Berschin, *Greek Letters and the Latin Middle Ages: From Jerome to Nicolas of Cusa,* trans. J. C. Frakes (Washington, 1988), 77–78.

43. Cassiodorus *Institutiones* I.25.2 (trans. p. 125): *tum si vos notitiae nobilis cura flammaverit, habetis Ptolomei codicem, qui sic omnia loca evidenter expressit, ut eum cunctarum regionum paene incolam fuisse iudicetis, eoque fiat ut uno loco positi, sicut monachos decet, animo percurratis quod aliquorum peregrinatio plurimo labore collegit.*

44. Ibid. II.6.1 (trans. pp. 197–98, modified by me): *. . . Varro, peritissimus Latinorum. . . . : tunc et dimensionem universae terrae probabili refert ratione collectam; ideoque factum est ut disciplina ipsa Geometria nomen acciperet. . . .*

of the circumferences of heaven and earth in stades," in a book that "briefly reveals many secrets of the philosophers."[45] This book, *De die natali,* was written in the third century and treated, among other things, geometry and land measurement.[46]

Cassiodorus did not seem to consider the descriptions of places the proper matter of geometry. At least, the rest of his chapter on geometry, in which he gives a brief description of the subject and recommends Boethius' translation of Euclid, does not include any descriptive material. Cassiodorus seems to connect geometry with land measurement—both as land surveying and calculations of the size of the earth—rather than with land descriptions.[47]

Thus, in various contexts (history, cosmography, geometry), Cassiodorus gives his monks a program of geographical studies, directed at better understanding the Bible. This rather ambitious program includes classical writings and covers much more than is necessary for simple identification of biblical place-names. We do not know, however, if this program ever became reality, either at Vivarium in Cassiodorus' time or in medieval schools later.[48]

---

45. Ibid., (trans. p. 198): *Unde Censorinus, in libro quem scripsit ad Quintum Cerellium, spatia ipsa caeli terraeque ambitum per numerum stadiorum distincta curiositate descripsit; quem si quis recensere voluerit, multa philosophorum mysteria brevi lectione cognoscit.*

46. On Censorinus' editions and transmission see R. H. Rouse and R. M. Thomson, "Censorinus," in *Texts and Transmissions: A Survey of the Latin Classics,* ed. L. D. Reynolds (Oxford, 1983), 48–50.

47. See also Cassiodorus *Variae* III.52, where Cassiodorus clearly connects geometry and the surveyor's art. B. L. Ullman has argued that agrimensorial tracts (tracts on land surveying) survived throughout the Middle Ages primarily as a source for geometrical material taught in the quadrivium, in "Geometry in the Mediaeval Quadrivium," in *Studi di bibliografia e di storia in onore di Tammaro de Marinis* (Verona, 1964), 4:263–85. The context of land surveying, closely related to geometry, could also accommodate certain geographical information. For instance, manuscripts containing tracts of Roman *agrimensores,* going back to late antiquity and copied throughout the Middle Ages, included local maps and plans of settlements: see O. A. W. Dilke, "Maps in the Treatises of Roman Land Surveyors," *Geographical Journal* 127 (1961): 417–26, and "Illustrations from Roman Surveyors' Manuals," *Imago mundi* 21 (1967): 9–29. Because agrimensorial tracts represent a separate group of technical sources only occasionally treating geographical matters, the connections between land surveying and geographical knowledge will not be treated in this book.

48. Pierre Riché (*Education and Culture,* 168–69) thinks it was never realized at Vivarium. James O'Donnell (*Cassiodorus,* 219–20) claims that the study of secular classics at Vivarium was far more limited than the *Institutions* suggest. To assess the situation in medieval schools, we need to ask what the manuscript transmission of Cassiodorus' *Institutions* can tell us about geographical studies. To my knowledge, this work has never been done.

Geographical Studies and *Physica*

Augustine's tradition to treat geographical studies as part of knowledge about nature, or *physica*—intended to serve the knowledge about God, or understanding of the Bible—further continued in the early Middle Ages in the works of Isidore of Seville (ca. 570–636). Following Augustine's program of "Christian doctrine," Isidore, in his *Etymologies,* an encyclopedia very influential throughout the Middle Ages, addressed all the subjects that Augustine considered necessary for a Christian to know. Contributing to the creation of the Christian corpus of knowledge, Isidore incorporated classical sciences into the Christian framework.[49] Borrowing Cassiodorus' definition of philosophy as "the knowledge of things human and divine," he offered two classifications of this knowledge. The first, tripartite division into *physica, ethica,* and *logica,* with the quadrivium included in physics, ultimately went back to the Stoic philosophy.[50] The second one was borrowed from Cassiodorus.[51] Even though neither of these divisions explicitly mentions geographical matters, they occupy a significant place in the *Etymologies* as a separate portion.

Isidore treats geographical material in two books, XIII–XIV, as part of the knowledge about the created world, *mundus,* which consists of "heaven, earth, sea, and all God's creations in them."[52] In the beginning of book XIII, he thus summarizes its contents: "some causes of the heaven

---

49. On Isidore's work as a response to Augustine's program see J. Fontaine, "Isidore de Seville et la mutation de l'encyclopédisme antique," in *La pensée encyclopédique au Moyen Age* (Neuchâtel, 1966), pp. 519–38, reprinted in J. Fontaine, *Tradition et actualité chez Isidore de Seville,* chap. 4 (London, 1988), esp. pp. 523–24. On Isidore and his influence see J. Fontaine, *Isidore of Seville et la culture classique dans l'Espagne wisigothique,* 2 vols. (Paris, 1959); A. E. Anspach, "Das Fortleben Isidors im VII. bis IX Jahrhundert," in *Miscellanea Isidoriana, Homenaje a S. Isidoro de Sevilla en el XIII centenario de su muerte* (Rome, 1936), 323–56; B. Bischoff, "Die europaeische Verbreitung der Werke Isidors von Sevilla," in *Isidoriana* (Leon, 1961), 317–44, reprinted in B. Bischoff, *Mittelalterliche Studien: Ausgewählte Aufsätze zur Schriftkunde und Literaturgeschichte* (Stuttgart, 1966), 1:171–94; J. Fontaine, "La diffusion carolingienne du *De natura rerum* d'Isidore de Seville d' après les manuscrits conservés en Italie," *Studi medievali,* 3d ser., 7 (1966): 108–27; J. M. Fernandez Caton, *Las Etimologias en la tradicion manuscrita medieval* (Leon, 1966).

50. Isidore, *Isidori Hispalensis episcopi Etymologiarum sive Originum libri XX,* ed. W. M. Lindsay, 2 vols. (Oxford, 1911), II.24.3: *Philosophiae species tripertita est: una naturalis, quae Graece Physica appellatur, in qua de naturae inquisitione disseritur . . . ;* II.24.4: *[Physicam] postmodum Plato in quattuor definitiones distribuit, id est Arithmeticam, Geometricam, Musicam, Astronomiam.* On Isidore's classification and its Stoic roots see J. A. Weisheipl, "Classification of the Sciences in Medieval Thought," *Mediaeval Studies* 27 (1965): 63–64.

51. Isidore *Etym.* II.24.10–16.

52. Ibid., XIII.1: *Mundus est caelum et terra, mare et quae in eis opera Dei.*

and the location of the lands and the spaces of the sea."[53] In Isidore's *Etymologies*, geographical material directly belongs to *physica*, the part of philosophy that treats "the inquisition of nature."[54]

Early medieval tracts titled *On the Nature of Things*, specially devoted to *physica* and written by Isidore of Seville and, later, by Bede (ca. 672–735), place geographical material in the same context. Both Isidore's tract and Bede's treat these matters very briefly, omitting the description of places and only offering a very general outline of the world.

Writing his treatise in order to explain "the nature and causes of things,"[55] Isidore begins with matters of the calendar and proceeds to describe the created world, *mundus*, consisting of heaven and the earth, thus introducing astronomical and geographical material. The geographical material includes a brief account of the ocean and seas, major rivers, and parts of the world.[56]

Bede, developing Isidore's tradition, devoted his *On the Nature of Things* to the created and visible world. Beginning with the account of the creation, he then describes the world *(mundus)*, which consists, in the words borrowed from Isidore, of heaven and earth. Bede describes it in the same order, beginning with astronomical questions and concluding with geographical ones.[57] Turning to Pliny the Elder to supplement information borrowed from Isidore, Bede mentions the shape of the earth, its climatic zones, and, in one paragraph, the three continents.

Carolingian scholars, following their predecessors' ideas about Christian knowledge and education,[58] placed geographical studies in the same context. Although occupying no special niche in any classification, geographical information continued to figure in ninth-century writings that summarized Christian knowledge in the spirit of Augustine's *doctrina Christiana*. Hrabanus Maurus, in his encyclopedia *On the Natures of Things*, composed between 842 and 847 as a reference tool for reading the

---

53. Ibid., XIII: . . . *quasdam caeli causas situsque terrarum et maris spatia.* . . .

54. Ibid., II.24.1: *una naturalis, quae Graece Physica appellatur, in qua de naturae inquisitione disseritur.* . . .

55. Isidore, *De natura rerum. Isidore de Séville, Traité de la nature*, ed. J. Fontaine (Bordeaux, 1960), *Praefatio* 1: . . . *et quaedam ex rerum natura uel causis a me tibi efflagitas suffragandum.*

56. Ibid., IX.1: *Mundus est uniuersitas omnis quae constat ex caelo et terra;* XL–XLVIII are the geographical chapters.

57. Bede, *De natura rerum*, ed. C. W. Jones, CCSL 123 (Turnhout, 1975), 172–234, III: *Mundus est uniuersitas omnis, quae constat ex caelo et terra* . . . ; XXXVIII–LI.

58. On Carolingian divisions of knowledge see P. Riché, *Les écoles et l'enseignement dans l'Occident chrétien de la fin du Ve siècle au milieu du XIe siècle* (Paris, 1979), 264.

Scriptures,[59] continued Isidore's tradition. Even though Hrabanus considerably changed the order of his encyclopedia compared to that of Isidore, his account of the created world follows Isidore's structure and puts geographical material in a similar context of *physica.*[60] Borrowing the contents of Isidore's geographical books, Hrabanus describes the earth—including its seas, rivers, and regions—in considerable detail.

John Scottus Eriugena (in the second half of the ninth century) included theoretical geographical material in his dialogue *On Natures.* Understanding the word *nature* in a wide sense that embraced both the Creator and the created world, John Scottus gave a synthesis of Christian knowledge, appropriately calling it *physiologia* (knowledge about nature).[61] In book III, devoted to the created world, he discussed some geographical questions, such as size and form of the earth. According to the traditional medieval classification of knowledge that John Scottus used, all these questions belonged to *physica,* the part of philosophy dealing with the created nature.[62]

Thus throughout the early Middle Ages, knowledge about nature provided a context for geographical studies. Continuing Augustine's tradition, Christian compendiums always included descriptions of the earth in their treatment of the created world, with various degrees of detail; used at school, these texts formed the basis of geographical studies.

---

59. Hrabanus Maurus, *Rabani Mauri de universo libri viginti duo,* PL 111, 9–614. Hereafter I refer to Hrabanus' work by its earlier, correct, title—*De rerum naturis.* On the date and title see E. Heyse, *Hrabanus Maurus' Enzyklopädie "De rerum naturis": Untersuchungen zu den Quellen und zur Methode der Kompilation* (Munich, 1969), 1–3. On Hrabanus' intention to create a reference tool see Hrabanus *De rerum naturis, Praefatio altera* 11–12D: *quo haberes ob commemorationem in paucis breviter adnotatum quod ante in multorum codicum amplitudine et facunda oratorum locutione disertum copiose legisti.*

60. See Hrabanus *De rerum naturis* IX on *mundus,* elements, atoms, etc.; X on chronology; XI–XIII on geography. Hrabanus defines philosophy at XV.1.416A: *Philosophia ergo est naturae inquisitio, rerum humanarum divinarumque cognitio.* . . . For Hrabanus on *physica* see XV.1.416B.

61. On the *Periphyseon* as a compendium of Christian knowledge see G. Schrimpf, "Die Sinnmitte von Periphyseon," in *Jean Scot Erigène et l'histoire de la philosophie: Colloque Internationale du Centre de la Recherches Scientifiques, Laon, 7–12 juillet 1975,* ed. R. Roques (Paris, 1977), 289–305. On *physiologia* see John Scottus *Periphyseon* IV.741C; for the edition of books I–IV see John Scottus, *Iohannis Scotti Eriugenae Periphyseon (De diuisione naturae),* vols. 1–3, ed. and trans. I. P. Sheldon-Williams and L. Bieler, and vol. 4, ed. E. A. Jeauneau with the assistance of M. A. Zier, trans. J. J. O'Meara and I. P. Sheldon-Williams (Dublin, 1968–95); for the edition of book V see John Scottus, *De divisione naturae,* PL 122, 439–1022. Page numbers in subsequent citations of this text refer to the parallel translation given in the 1968–95 edition.

62. John Scottus *Periphyseon* III.3.629B: *Est enim physica naturarum sensibus intellectibusque succumbentium naturalis scientia.* . . . See also III.29.705B.

Geographical Studies and the Quadrivium

Another context for geographical studies at school was provided by a non-Christian authority. In the fifth century, Martianus Capella wrote his encyclopedia *On the Marriage of Philology and Mercury,* which, along with later works by Boethius and Cassiodorus, established the canon of the seven liberal arts in education. Martianus also established a solid educational context for geographical studies, by placing geography within the quadrivium and treating it as part of geometry.[63]

In book VI, Geometry, one of the seven personified liberal arts in the story, introduces herself to the guests at the celestial wedding of Philology and Mercury, claiming direct connection to knowledge about the earth. She says: "I am called Geometry because I have often traversed and measured out the earth, and I could offer calculations and proofs for its shape, size, position, regions, and dimensions. There is no portion of the earth's surface that I could not describe from memory."[64] Thus, the self-definition of Geometry includes both the measurement of the earth and the description of its surface and provides the first justification for her later, extensive geographical discourse.

Another, literary justification is provided by the unfamiliarity of the celestial wedding guests with the earth. The guests—divinities of various statuses, many of whom dwell in celestial regions and "had never trodden upon the earth"—encourage Geometry to begin with the description of the earth and to only then turn to "the other precepts of her discipline," that is, geometry proper.[65]

This literary justification wears thin by the end of book VI, when the guests get tired of Geometry's long and tedious digression and begin to complain: Why do they have to listen to this "pitiless boor with coarse limbs," this creature, so dusty, tough, and peasantlike that one could take her for a man? Why do they have to listen to all this geography that has

---

63. Martianus Capella *De nuptiis* VI. On Martianus' date see Danuta Shanzer's review of the problem in *A Philosophical and Literary Commentary on Martianus Capella's "De nuptiis Philologiae et Mercurii" Book 1* (Berkeley, Los Angeles, and London, 1986), 5–8, and her arguments (10–28) for the later date of the 470s rather than the earlier one of 410–39.

64. Martianus Capella *De nuptiis* VI.588 (trans. p. 220): *Geometria dicor, quod permeatam crebro admensamque tellurem eiusque figuram, magnitudinem, locum, partis et stadia possim cum suis rationibus explicare, neque ulla sit in totius terrae diuersitate partitio, quam non memoris cursu descriptionis absoluam.*

65. Ibid., VI.589 (trans. p. 220): *. . . quoniam fuerant in deorum senatu quamplures, qui neque orti terris essent neque ipsi umquam dicerent se calcasse tellurem, . . . Geometria primum iubetur ac demum cetera astruendae praecepta artis aperire.*

almost nothing to do with the subject Geometry had promised to talk about?[66]

It is tempting but very risky to take the displeasure of the wedding guests as an indication of the attitudes of the real audience in the fifth century. It is first and foremost a literary device: Grammar, Dialectic, and Rhetoric, who seem to say nothing outside their respective fields, are also ordered to stop their lengthy discourse.[67] It appears that the guests object more strongly to the length of the geographical discourse than to the whole idea of relating it to geometry, but Martianus' text itself gives us no clear indication of how his approach might have been perceived by his contemporaries or why he did it at all.

These tantalizing questions always appear in Martianus' studies because his book is the first and only one of the extant Latin texts that gives geographical material such a prominent place within geometry. Friedrich Ritschl and Manfred Simon directed attention away from Martianus by saying that he was only following the tradition of the Roman encyclopedic writer Marcus Terentius Varro.[68] However, since we do not have the text of Varro's *Disciplines,* this remains a pure conjecture.

William Stahl, who is in general very skeptical about Martianus' "scientific" achievements, suggested that Martianus, not having enough material or enough of a handle on geometry proper, decided to fill out book VI with geographical material, choosing it because of the popularity of his two chief sources, Pliny and Solinus.[69] Brigitte Englisch, answering this argument, pointed out Martianus' more solid scholarly reasons. She argued that Martianus, realizing the tight connection between geography and geometry demonstrated by the methods of land surveying, was building on geometrical constructs but, at the same time, did not want to simply mix the two subjects together, so he logically separated them by an allegorical interlude, which never happens in other books.[70]

Although we cannot be sure about Martianus' motives or predecessors

---

66. Ibid., VI.704–5 (trans. pp. 263–64).
67. Ibid., III.326, IV.423, V.565.
68. F. Ritschl, *De M. Terentii Varronis Disciplinarum libris commentarius,* in F. Ritschl, *Kleine philologische Schriften (Opuscula philologica)* (Leipzig, 1877), 3:352–402, esp. 387–97; M. Simon, "Zur Abhängigkeit spätrömischer Enzyklopädien der Artes Liberales von Varros Disciplinarum libri," *Philologus* 110 (1966): 88–101.
69. W. Stahl, "The Quadrivium of Martianus Capella: Latin Traditions in the Mathematical Sciences, 50 B.C.–A.D. 1250," in *Martianus Capella and the Seven Liberal Arts* (New York, 1977), 1:128–29.
70. B. Englisch, *Die Artes liberales im frühen Mittelalter* (Stuttgart, 1994), 155.

in this particular matter, it seems certain that he had started a new tradition in education with respect to geographical studies. Martianus' text became a popular manual in the Middle Ages, and those schools that used it studied geographical material as part of the liberal arts curriculum.

We know very little about the use of Martianus' book before the ninth century. We do not even know if he meant it for the schoolroom.[71] It may have been used at school between the time it was written and A.D. 534, when Securus Felix, a rhetor, emended the text "from most corrupt exemplars." Otherwise, how did these exemplars survive? Securus Felix may have used Martianus in his teaching (he mentions at least one of his disciples, Deuterius the *scholasticus,* who helped him to emend the text).[72] If Securus Felix used the entire book, geographical material—and, probably, the rest of the quadrivium—may have served the rhetorical education.

Gregory of Tours (539–94) testifies to Martianus' use in sixth-century schools, when he connects Martianus' name to instruction in the seven liberal arts. Among other arts, he refers to geometry, which teaches "to calculate measurements of lands and lines."[73] Gregory seems to emphasize the measurement of the earth, omitting the description of places, but his evidence is too general to make any conclusions about geographical studies.

Our earliest detailed evidence about Martianus' geography at school comes from the second half of the ninth century, when several Carolingian schools began to use Martianus' encyclopedia as a textbook. At that time, the fact that the book on geometry treated mainly geography did not seem strange at all. Unlike Martianus' wedding guests, ninth- and tenth-century scholars seemed to take Geometry's geographical digression for granted. In commenting on this passage, they never address the relations of the two

---

71. Danuta Shanzer (*Philosophical and Literary Commentary,* 51–52) suggests that Martianus meant his work as a philosophical dialogue in the Platonic tradition or as a revelational text.

72. The following is recorded in many manuscripts: *Securus Melior Felix, vir spectabilis, comes consistorii, rhetor urbis Romae, ex mendosissimis exemplaribus emendabam, contra legente Deuterio scolastico discipulo meo, Romae, ad Portam Capenam, consulatu Paulini viri clarissimi, sub die nonarum martiarum, Christo adjuvante.* See C. Leonardi, "I codici di Marziano Capella," *Aevum* 33 (1959): 443–89, 34 (1960): 1–99, 411–524 esp. 1959, 446–47; C. Lutz, "Martianus Capella," in *Catalogus translationum et commentariorum: Medieval and Renaissance Latin Translations and Commentaries, Annotated Lists and Guides,* ed. P. O. Kristeller and F. E. Cranz (Washington, 1971), 2:367–81, esp. 368.

73. Gregory of Tours, *Historiae,* ed B. Krusch, MGH SRM I (Hanover, 1951), X.31: *in geometricis terrarum linearumque mensuras colligere.* M. Laistner (*Thought and Letters,* 129) suggests that by Gregory's time, Martianus' *De nuptiis* had become a standard treatise on the liberal arts; see also reservations expressed in Stahl, "Quadrivium," 59 n. 13.

subjects.[74] According to them, the wedding guests were irritated by the length of the speech, not by its contents: "for their pleasure [from Geometry's discourse] turned into boredom because of her long speech."[75]

In the early medieval system of education and knowledge, Martianus' tradition seems to be the only one that explicitly connected geographical studies with geometry and secured their place within the quadrivium. While measurement of the earth occasionally figures in geometrical texts, no other tradition except Martianus' includes the descriptive material.

Though Boethius and Cassiodorus, along with Martianus, shaped early medieval ideas about education and knowledge and about the place of the liberal arts within them, neither gave geographical studies such prominence. It also appears that neither of them was directly familiar with Martianus' tradition.[76] Boethius does not mention geographical questions either in his classification of knowledge—with a tripartite division of philosophy into physics, mathematics, and theology—which was to become standard for the Middle Ages, or in his extant handbooks on the liberal arts.[77] Cassiodorus, suggesting several contexts for geographical studies, explicitly connected none to the liberal arts.

Isidore does not treat geography as part of the quadrivium. In book III, writing about geometry, he does not include description of places, and he only briefly mentions land measurement. He does it in the same context, if not in the same words, as Cassiodorus had done—explaining the origins of geometry: "as this discipline begins from the measurement of the earth, so from the beginning it used its name. For geometry is called after 'earth' and 'measurement.'"[78] However, unlike Cassiodorus, Isidore does not emphasize the use of geometry for land measurement. For Isidore, geography seems to lie outside the arts of the quadrivium, even though both the

---

74. This is true in the commentaries that I have studied; further investigation may change the picture.

75. Remigius *Commentum* VI.351.7: *quia voluptas illarum in tedium vertebatur per illius narrationem prolixam.*

76. There is no evidence of any Martianean influence in Boethius' quadrivial manuals; see the discussion and references in Stahl, "Quadrivium," 57–58, and Shanzer, *Philosophical and Literary Commentary,* 11–12. Cassiodorus (*Institutiones* II.2.17, II.3.20) mentions *De nuptiis* but complains that he could not obtain a copy.

77. On Boethius' classification see J. Mariétan, *Problème de la classification des sciences d'Aristote à s. Thomas* (Paris, 1901), 63–71, and Weisheipl, "Classification," 60–61; for extant handbooks by Boethius on arithmetic and geometry see Boethius *De institutione arithmetica,* ed. G. Friedlein (Leipzig, 1867).

78. Isidore *Etym.* III.10.3: . . . .*quia ex terrae dimensione haec disciplina coepit, ex initio sui et nomen servavit. Nam geometria de terra et de mensura nuncupata est.*

former and the latter belong to *physica,* with the mathematical arts forming its four divisions.⁷⁹

It is hard to tell whether Hrabanus Maurus considered geographical material part of the quadrivium. In his encyclopedia, he only gives definitions of the liberal arts, devoting no special attention to them. However, his definition of geometry may include the measurement of the earth: "Geometry is the discipline of measuring the distances between [or "sizes of"] places and the magnitudes of bodies."⁸⁰

John Scottus related the earth measurement to geometrical context, despite his rather narrow and technical definition of geometry.⁸¹ When he discussed the size of the earth and explained Eratosthenes' method of calculating the earth's circumference, he drew his material from Martianus Capella and, like Martianus, connected it to geometry. He referred to the measurements of the earth as geometrical and distinguished their methods from those of land surveying. While land surveyors use tools, such as geometrical rods, the circumference of the earth is calculated "by logical argument alone, that is, by means of sundials."⁸² These geometrical measurements are treated in geographical books and belong to "geographical speculation."⁸³ John Scottus does not give descriptions of places, but we have no way to find out whether he did not relate these things to "geographical speculation" or simply omitted them because they did not fit in his plan.

The same trend to include in geometry the measurement but not the actual description of the earth continued in the tenth century. The anonymous tenth-century textbook on geometry described at length Eratosthenes' measurement of the earth, borrowing the material, if not the exact words, from Martianus Capella.⁸⁴

---

79. Ibid., II.24.4: . . . .*quam [physicam] postmodum Plato in quattuor definitiones distribuit, id est Arithmeticam, Geometricam, Musicam, Astronomiam.*

80. Hrabanus Maurus *De rerum naturis* XV.1.413D: *Geometria est disciplina mensurandi spatia locorum, et magnitudines corporum.*

81. John Scottus *Periphyseon* I.475A: *Geometrica est planarum figurarum solidarumque spatia superficiesque sagaci mentis intuitu considerans disciplina.*

82. Ibid., III.716 C: . . . *doctissimus in omni geometrica et astrologica supputatione Eratosthenes* . . . ; III.725B: *virga . . . geometrica;* III.725C, p. 269: *Non . . . pedibus neque radiis, sed sola rationis argumentatione, horologiis uidelicet.* . . .

83. Ibid., III.725D: *Dimensiones geometricas talibus argumentis ita primo repertas non contradixerim* . . . ; III.719A: *Si autem quaeras cur et Plinius Secundus et Ptolomeus in geografico suo, ut Martianus scribit* . . . ; III.726A: *Hinc est quod et Plinius amplitudinem terrae ducta rationabili linea geografica speculatione per cacumina altissimorum montium mensurari existimat.*

84. *Geometria incerti auctoris,* in *Gerberti postea Silvestri II papae Opera Mathematica (972–1003),* ed. N. Bubnov (Hildesheim, 1963), 310–65, IV.60.

Even though geometrical texts did not normally include geographical descriptions, manuscripts testify that the two subjects were perceived as related: geometrical and geographical texts were sometimes transmitted in the same codices. For instance, in a volume from Reichenau (before 842), cosmographical texts by Julius Honorius, Aethicus, and Jerome are transmitted along with Boethius' *Geometry*.[85]

Thus, only Martianus Capella and the ninth- and tenth-century schools that used his encyclopedia as a textbook extensively treated geographical material in the context of geometry, including both theoretical questions, such as earth measurement, and the detailed descriptions of places. Other authors, only relating earth measurement to geometry, never seem to have thought that the description of the earth—including its continents and regions—belonged to this discipline.

Geography Per Se

Although geography lacked a definite disciplinary status in the early Middle Ages, geographical knowledge was considered important and self-sufficient enough to become a subject of special treatises—school texts, more advanced scholarly works, and even parodies. Geographical texts transmitted together sometimes formed manuscript collections specially devoted to knowledge about the earth. Examples are MSS Paris, BN lat. 4806, produced at Reims in the middle or the third quarter of the ninth century, and Dresden, Sächsische Landesbibliothek Dc 182 I, written in the late ninth or early tenth century and destroyed during the Second World War.[86] These manuscripts contain the eighth-century *Cosmography* of Pseudo-Aethicus; the tract *On the Measurement of the Earth,* written in 825 by Dicuil, an Irish teacher at the Carolingian court; and late antique topographical writings.[87]

The perception of geographical knowledge as to a certain degree self-

---

85. Becker, *Catalogi,* 10, 2; on this see also Ullman, "Geometry," 277–78, and M. Folkerts, "The Importance of the Pseudo-Boethian *Geometria,*" in *Boethius and the Liberal Arts* (Bern, 1981), 187–209, esp. 198.

86. On both manuscripts see L. Bieler, "The Text Tradition of Dicuil's *Liber de mensura orbis terrae," Proceedings of the Royal Irish Academy* 64 C 1 (1965): 1–31, esp. 3–5. Both codices are related to the now lost manuscript of the late ninth century or early tenth century from Speyer that contained a larger number of geographical and topographical texts (see ibid., 1–3).

87. The late antique topographical writings are *Itinerarium Antonini, Septem montes urbis Romae,* and *De aquis urbis Romae;* see Bieler, "Text Tradition," 2.

sufficient is also demonstrated by the fact that authors of special treatises usually said nothing or very little about the intellectual or disciplinary context of their writings. The existing ideas about the hierarchy of knowledge and the place of geography within it did not need to be verbalized every time and existed on the level of tacit assumptions.

Thus when the authors of geographical tracts declared that they wished to describe the earth *(orbis terrarum)*, their early medieval audience would most likely perceive this subject as part of *physica*, knowledge about the created world. Dicuil, who very briefly defined the subject of his book as "the measurement of the provinces of the earth,"[88] seems to have regarded it as pertaining to the knowledge about the created, corporeal, visible world. Only once, in passing, does he mention it. Referring to "the seven things" described in one of his sources—seas, islands, mountains, provinces, cities, rivers, and nations—he later defends his use of the word *res* (things) in relation to these corporeal and visible matters.[89]

Dicuil's tract, recording dimensions and distances of continents, seas, mountains, rivers, and provinces, could also have been used in the context of geometry; in fact, the very title of Dicuil's work evokes the traditional late classical and early medieval definition of this discipline, "the measurement of the earth."[90] Dicuil, who may have taught at the palace school of Louis the Pious, composed a number of other school texts for the study of the liberal arts.[91] His *On the First Syllable* and the lost *Letter on the Ten Questions of the Grammatical Art* were intended for rhetoric and grammar. Mathematical arts of the quadrivium would form an appropriate setting for Dicuil's interest in measurements, also attested in his tracts on weights and measures and on calendrical calculations.[92]

Only rarely would an author refer to a wider context of geographical

---

88. Dicuil *De mensura, Prologus* 1: . . . *liber de mensura provinciarum orbis terrae.* On the date see Tierney's introduction in Dicuil, *Dicuili liber,* 17.

89. In *De mensura* VIII.26–29 Dicuil talks about "the seven matters": *De septem rebus sequentibus in Cosmographia haec scripta sunt: Orientalis pars [orbis terrarum] habet maria VIII, insulas VIIII, montes VII . . . Meridiana pars habet maria II, insulas XVI, montes VI. . . .* Further, in VIII.31 he defends his use of *res* by drawing examples from grammarians: *Sed ne litterator reprehendat quod corporales et uisibiles hic dixi. . . .*

90. This definition was used by Cassiodorus, Martianus Capella, Gregory of Tours, Isidore of Seville, and Hrabanus Maurus and by John Scottus Eriugena in his commentary on Martianus' encyclopedia; see Gautier Dalché, "Tradition," 139–40.

91. For biographical information about Dicuil see Tierney's introduction in Dicuil, *Dicuili liber,* 11–17.

92. Dicuil's *Computus* is edited in M. Esposito, "An Unpublished Astronomical Treatise by the Irish Monk Dicuil," *Proceedings of the Royal Irish Academy,* 26 C (1907): 381–445; on the *Epistula censuum* see Tierney's introduction in Dicuil, *Dicuili liber,* 14–16.

knowledge, as did an anonymous author from Ravenna, Anonymus Ravennatis, who wrote his *Cosmography* in the early eighth century. He regarded his treatise, which described "the whole world and the places where various nations live,"[93] as part of the knowledge about the created world, ultimately meant to help the student understand the Creator. Never saying this directly, he implies it by arranging biblical quotations in the beginning. Seeing his work as a service to God, inspired and watched over by him, Anonymus Ravennatis wishes to belong to "those who perform their service under the order of God, engage in observing his law, and desire to delight in the good of a blessed life and infinite glory."[94] These considerations, as Anonymus says, along with the divinely inspired request of a certain Brother Odo, served him as an incentive to write a description of the world: "On account of this, my dearest brother, since you, possessed by the divine inspiration, compel me to describe the world in detail . . ."[95] This labor, justified as an answer to the divinely inspired request, is going to serve the highest goal of Christian learning—to understand God and his creation. Again by means of biblical quotations, Anonymus Ravennatis refers to the miracle of the creation of the earth, its dimensions, and its beauty and asserts that a human mind is able to give tribute to the divine work: "For it is possible for human understanding: with God's help, I have read the books of many philosophers as much as I could."[96]

His account of the earth, combining late Roman and biblical traditions, reflects the early medieval hierarchy of knowledge. He organizes his account around fundamental biblical postulates, using the methods of biblical exegesis. The Scripture helps him to justify his plan to describe the earth by dividing it into twenty-four segments, corresponding to the num-

---

93. Anonymus Ravennatis, *Cosmographia*, in *Itineraria Romana*, vol. 1, ed. J. Schnetz (Stuttgart, 1990), I.1, p. 1 (page numbers in subsequent citations of Anonymus Ravennatis *Cosmographia* refer to this edition): . . . *totum mundum diversarumque gentium habitationes.* . . . On Anonymus see L. Dillemann, *La Cosmographie du Ravennate*, ed. Y. Janvier (Brussels, 1997); J. Schnetz, *Untersuchungen zum Geographen von Ravenna* (Munich, 1919), and *Untersuchungen über die Quellen der Kosmographie des anonymen Geographen von Ravenna* (Munich, 1942). For more bibliography see F. Staab, "Ostrogothic Geographers at the Court of Theodoric the Great: A Study of Some Sources of the Anonymous Cosmographer of Ravenna," *Viator* 7 (1976): 27–58.

94. Anonymus Ravennatis *Cosmographia* I.1, p. 1: *Sub dei qui militant imperio, eius legem observare iniant et beate uite atque infinite glorie perfrui bono desiderant.* . . .

95. Ibid., I.1, p. 1: *Quam ob rem, o mi frater carissime, postquam divina aspiratione praeditus me compelleres, ut ego per pallidines subtilius tibi indicem mundum.* . . .

96. Ibid., I.1, p. 1: *sed tantum cum propheta clamemus "quam magnificata sunt opera tua, domine! omnia in sapientia fecisti." nam quod apud humanum sensum possibile est: multorum phylosophorum relegi libros Christo iuvante in quantum valeo.*

ber of hours in a day—an idea probably going back to the Roman practice of land surveying.[97] Anonymous acknowledges that God, in creating the luminous bodies in the sky, designed it so that they could serve for the reckoning of time. He maintains that just as the sun, performing its circle around the earth, designates hours of the day and night on the circle of a sundial, "we can consider and designate in detail, with Christ's help, the lands of all the nations, placed around the great circuit of the impassable ocean."[98] Within this structure, Anonymus also uses another one, the traditional classical division into three continents, which he complements by biblical information—the names of the sons of Noah who settled in each continent.

Thus demonstrating the connections between geographical knowledge and the Bible, Anonymus Ravennatis gives no specifics about how he envisioned the practical use or disciplinary context of his tract. The meager information that Anonymus Ravennatis gives about "Brother Odo"—that he had asked Anonymous to write the *Cosmography*—may suggest that Odo was a disciple or a fellow scholar and that the tract may have been written with an educational goal in mind.

That geographical tracts were popular and widely studied in learned circles is also demonstrated by an attempt to parody them. An anonymous author of an eighth-century *Cosmography,* assuming the venerable name of Jerome, declares that he had translated the work of the otherwise unknown philosopher Aethicus Ister.[99] The author claims that Aethicus, addressing many questions omitted by Moses in the Old Testament,[100] had

---

97. On Anonymus' plan and its origin see Schnetz, *Untersuchungen über die Quellen,* 12–14; on its inconsistencies see Staab, "Ostrogothic Geographers," 31–32.

98. Anonymus Ravennatis *Cosmographia* I.1, p. 2: *ergo dum sol totam diem per meridianum marginem potentissime iussu factoris exambulat et unamquamque horam diei verno tempore, tanquam horologium per ordinem suum, per occursum designat, possumus arbitrari universarum gentium patrias per magnum circuitum intransmeabilis oceani litore positas et Christo nobis auxiliante subtilius designare.*

99. Aethicus Ister, *Die Kosmographie des Aethicus,* ed. O. Prinz (Munich, 1993); page numbers in subsequent citations of Aethicus Ister *Cosmographia* refer to this edition. This text has attracted much scholarly attention—see the survey and bibliography in Prinz's introduction to *Die Kosmographie,* 1–84. On this work as a parody see K. Hillkowitz, *Zur Kosmographie des Aethicus,* vol. 1 (Cologne, 1934), vol. 2 (Frankfurt am Main, 1973), 1:73, and, particularly, H. Löwe, "Ein literarische Widersacher des Bonifatius: Virgil von Salzburg und die Kosmographie des Aethicus Ister," *Abhandlungen der Akademie der Wissenschaften und der Literatur in Mainz, Geistes- und sozialwissenshaftliches Klasse* 11 (1951): 903–83, and H. Löwe, "Salzburg als Zentrum literarischen Schaffens im 8. Jahrhundert," *Mitteilungen der Gesellschaft für Salzburger Landeskunde* 115 (1975): 99–143.

100. Aethicus Ister *Cosmographia,* p. 88: . . . *Aethicus iste chosmografus tam difficilia appetisse didiceret, quaeque et Moyses et vetus historia in enarrando distulit.* . . .

drawn much of his material from his own travels. The *Cosmography* begins, in the tradition of the tracts titled *On the Nature of Things,* with the creation of the world. After discussing the position of the earth in the universe, the creation of Paradise, and the angels and the luminous bodies in the sky, the author turns to the description of the earth. He describes the regions both known and unknown from other sources, particularly emphasizing the latter. He talks about things, never seen by anybody, that Aethicus had reported in his *Cosmography:* remote northern islands, people living near the Caspian Sea, wondrous animals.[101]

Scholars who have studied this complicated text, which is not always clearly written, have established that much of it represents a mystification, beginning with the name of Jerome, the supposed translator. The famous father of the church who had produced the Latin translation of the Bible could not have also translated the work of Aethicus, for the latest source used in the *Cosmography* dates from the eighth century. The anonymous author, however, knew Jerome's works very well and used this knowledge, cleverly constructing the narrator's persona.[102] In the same way, borrowing from the learned tradition, he cleverly constructed names of places and people. For example, the name of a remote northern island seen only by Aethicus, Rifargica, or Riffarrica, was constructed by analogy with the Riphean Mountains, which, according to Isidore, were located in Germany.[103] The author often put his newly invented place-names in well-known regions, combining them with information drawn from respectable and familiar sources, such as Isidore's *Etymologies.*[104]

The patterns of this cosmographer's work have made many scholars conclude that much of his information, including the unknown place-names, was invented.[105] Even if the failure to identify many of his place-names, taken by itself, does not necessarily mean that they are fictitious,

---

101. Ibid., pp. 105–6, 130–37, 150ff.

102. On the time of composition see H. Löwe, "Die 'Vacetae insolae' und die Entstehungszeit der Kosmographie des Aethicus Ister," *Deutsches Archiv für Erforschung des Mittelalters* 31 (1975): 1–16; on the "Jerome" fiction see M. W. Herren, "Wozu diente die Fälschung der Kosmographie des Aethicus?" in *Lateinische Kultur im VIII. Jahrhundert: Traube-Gedenkschrift,* ed. A. Lehner and W. Berschin (St. Ottilien, 1989), 145–59.

103. Aethicus Ister *Cosmographia,* pp. 105 and 130; see the explanation in K. Hillkowitz, *Zur Kosmographie des Aethicus,* 2:143, n. 228; for other numerous examples see ibid.

104. On the invented place-names see Prinz's introduction to Aethicus Ister, *Die Kosmographie,* 22; on Isidore as the main source see Hillkowitz, *Zur Kosmographie,* 1:67–68, 2:143–45.

105. See Löwe, "Ein literarischer Widersacher," 94; Prinz's introduction to Aethicus Ister, *Die Kosmographie,* 22.

the strong emphasis of this cosmography on the unknown supports this conclusion. The way the author concentrates on Aethicus' personal experience, unsupported by any authority, seems an intentional mocking reversal of the contemporary scholarly procedure.[106]

In order for the *Cosmography* to have been intended as a parody on geographical tradition, there must have been a developed idea of what this tradition was like. The author of the *Cosmography* displayed his mastery of current geographical knowledge to such an extent that he could mock it. Even though his medieval audience seems to have largely missed the point,[107] his very intention testifies to the importance and development of geographical knowledge by the eighth century.

Lacking a definite name, theoretical geographical knowledge also lacked a definite disciplinary status. In late antiquity and the early Middle Ages, the material that we call geographical may have belonged to various contexts. In historical writings, it traditionally occupied an important place, being perceived as tightly connected to history. In encyclopedias and tracts titled *On the Nature of Things,* it constituted part of the knowledge about the created world, belonging to *physica.* Sometimes it was included in the quadrivium as part of geometry. This uncertain position of geographical studies within the medieval system of knowledge and education resembled the status of history, which also had no particular place in medieval divisions of knowledge or medieval curriculum, being sometimes perceived as part of grammar, or rhetoric, or calendrical studies.[108]

Even though geographical knowledge was not represented by an independent discipline, it achieved a certain self-sufficiency as a subject. Tracts specially devoted to geography, often without mentioning this directly, were perceived as treating part of the knowledge about the created world.

In these contexts, geographical knowledge, like all secular knowledge about nature and the world, was meant to serve spiritual knowledge about

---

106. This subject will be addressed in more detail in chap. 5.

107. In library catalogues it was sometimes listed among Jerome's works, and some maps used its information; see M. Zeiler, "'Quicumque aut quilibet sapiens Aethicum aut Mantuanum legerit' Muß der Name des Verfassers der Kosmographie wirklich 'in geheimnisvolles Dunkel gehüllt bleiben'?" *Wiener Studien* 104 (1991):183–207, esp. 184 and n. 83.

108. For a discussion and bibliography see H. W. Goetz, "Die 'Geschichte' im Wissenschaftssystem des Mittelalters," chap. 12 in F.-J. Schmale, *Funktion und Formen mittelalterlicher Geschichtsschreibung: Eine Einführung, mit einem Beitrag von Hans-Werner Goetz* (Darmstadt, 1985). I thank one of my anonymous readers for bringing this point to my attention.

God. By providing information for understanding the Scripture, geographical studies eventually served biblical exegesis. The latter, receiving and incorporating the material supplied by secular sciences and thus reflecting their ideas, in its turn provided them with paradigms that shaped their development.

CHAPTER 2

Exegesis and Geographical Tradition

Early medieval scholars perceived the Bible as both the source and the focus of all knowledge and studied it in numerous exegetical works. Paradigms created in exegesis penetrated many fields of early medieval learning and shaped their perspective. In history, to take only one instance, the course of human events received a whole new dimension. It was perceived as reflecting the divine plan and could be used for its interpretation. The connection between human history and God's design thus worked both ways; and methods borrowed from exegesis served to explain both.[1]

In the same way, biblical ideas about the world, reflected and explained in exegesis, were bound to influence the perception and description of the earth. Although historians have often stated this and studied certain aspects of this influence,[2] the relations between biblical exegesis and early medieval geographical tradition have never been fully explored. My aim here is to consider two questions that may serve as prolegomena to this

---

1. On the connections between medieval history writing and exegesis see Schmale, *Funktion*, esp. chaps. 7 and 12. On specific early medieval examples see R. W. Hanning, *The Vision of History in Early Britain, From Gildas to Geoffrey of Monmouth* (New York and London, 1966); R. D. Ray, "Bede, the Exegete, as Historian," in *Famulus Christi: Essays in Commemoration of the Thirteenth Centenary of the Birth of the Venerable Bede,* ed. G. Bonner (London, 1976), 125–40; C. B. Kendall, "Imitation and the Venerable Bede's *Historia Ecclesiastica,*" in *Saints, Scholars, and Heroes: Studies in Medieval Culture in Honor of Charles W. Jones,* ed. M. H. King and W. M. Stevens (Collegeville, 1979), 1:161–90.

2. All general histories of medieval geography say this: see Beazley, *Dawn,* 1:40–44; Kimble, *Geography,* 19–20. The questions best represented in literature are those involving the terrestrial Paradise and the settlement of Noah's sons: in addition to the works just cited, see, on the first question, A. Graf, *La leggenda del paradiso terrestre* (Turin, 1878); H. R. Patch, *The Other World according to Descriptions in Medieval Literature* (Cambridge, 1950); R. R. Grimm, *Paradisus coelestis, Paradisus terrestris: Zur Auslegungsgeschichte des Paradieses im Abendland bis um 1200* (Munich, 1977). On the second question, see D. Hay, *Europe: The Emergence of an Idea* (Edinburgh, 1957).

study: What were the implications of the biblical and exegetical perception of the earth for geographical descriptions? And how did the relation between exegesis and geographical tradition work? To study these questions, I will turn to early medieval exegetical texts that comment on several biblical passages and that I have found particularly revealing.[3]

How Can You See the Earth?

How did early medieval people perceive geographical space? How could they visualize the earth or large parts or it, such as whole countries, before the age of space and air travel? Two episodes in the Bible, describing two ways of perceiving the earth, attracted exegetical interpretations that give us an insight into these questions.

> "And the Devil took him up, and showed him all the kingdoms of the world in a moment of time. . . ."

In the first episode,[4] described in two Gospels (Matt. 4:8 and Luke 4:5), the devil takes Christ to the top of the mountain, shows him all the kingdoms of the world, and promises him their glory. The Gospels do not seem concerned with the question of how the whole world, with its kingdoms, could be perceived in one moment. Latin exegetes did not often ask this question either.[5] Most of those who explain this passage at all treat it allegorically.[6]

---

3. Recently this method of working with medieval biblical exegesis, a category of sources rarely studied except for the history of theology, has been successfully used to answer various questions of cultural and economic history. See J. Cohen, *"Be Fertile and Increase, Fill the Earth and Master It": The Ancient and Medieval Career of a Biblical Text* (Ithaca and London, 1992) and S. Epstein, "The Theory and Practice of the Just Wage," *Journal of Medieval History* 17 (1991): 53–69. On geographical questions see Hay, *Europe.* On astronomy see S. McCluskey, *Astronomies and Cultures in Early Medieval Europe* (Cambridge, 1998), 31–38.

4. Luke 4:5 Revised Standard Version. The idea to choose this episode was suggested to me by Beryl Smalley's brief mention of commentaries on it, without exact references, in "The Bible in Medieval Schools," in *The Cambridge History of the Bible,* vol. 2, *The West from the Fathers to the Reformation,* ed. G. W. H. Lampe (Cambridge, 1969), 211.

5. I have examined the texts, chronologically going as far as the tenth century, using the following databases: Patrologia Latina database (Alexandria, VA, 1995) and Cetedoc Library of Christian Latin texts [computer file] (Turnhout, 1991–).

6. On the methods of interpretation of the Bible, ultimately representing either literal or allegorical treatment of the text, as well as on other questions of medieval exegesis see B. Smalley, *The Study of the Bible in the Middle Ages* (Oxford, 1941); H. de Lubac, *Exégèse médiévale: Les quatres sens de l'Ecriture,* 4 vols. (Paris, 1959–65); the 3 vols. of *The Cambridge History of the Bible.* For the allegorical treatment of the passage in question see, for instance: Ambrose *Expositio Evangelii secundum Lucam,* ed. M. Adriaen, CCSL 14 (Turnhout, 1957),

Only two commentaries known to me, both produced in the second half of the ninth century, turn to the problem of seeing the whole world at once, explaining it literally and in a similar way, although not in the same words.[7]

A Carolingian scholar, Christian of Stavelot, or Christian Druthmar, offers two explanations in his commentary on Matthew written shortly after 864.[8] According to the first one, Christ could have allowed the devil "to show him the whole world in one vision, in the same way as, according to what we read, the blessed Benedict saw it." According to the second explanation, the devil in reality "showed him only a part and promised to give him the rest"; thus, the Scripture, as it often does, uses here a figure of speech called "synechdoche," only naming a part while meaning the whole.[9]

Christian's first explanation will concern us here, because of its wider implications for the perception and representation of the earth. In it, Christian compares Christ's vision to that of St. Benedict, described by Gregory the Great (590–604) in his *Dialogues*. One night, when St. Benedict was praying by the window, he saw light coming from above and dispelling the darkness. Gregory reports, "A most miraculous thing followed during this contemplation, for, as he [St. Benedict] himself recounted, the whole world, as if collected under a sole ray of the sun, was brought before his eyes."[10] Thus, Christian of Stavelot seems to think that the only way to perceive the whole world at once, even for Christ, is a contemplative vision like that of St. Benedict.

1–400, esp. 115–16; Jerome *Commentariorum in Mattheum libri IV*, ed. D. Hurst and M. Adriaen, CCSL 77 (Turnhout, 1969), 22; Hrabanus Maurus *Commentariorum in Matthaeum libri octo*, PL 107, 783 D-784 A; Smaragd *Collectiones in Epistolas et Evangelia*, PL 102, 122–23.

    7. I could not find the sources for either explanation among the edited commentaries.

    8. On Christian and his Matthew commentary see his letter of ca. 865–70 where he mentions the composition of his commentary: *Epistolae variorum* 24, pp. 177–78, ed. E. Dümmler, MGH Epistolarum 6, 127–206, (Munich, 1978); F. Stegmüller, *Repertorium Biblicum Medii Aevi* (Madrid, 1950), 2:239; *Lexikon für Theologie und Kirche*, ed. M. Buchberger, (Freiburg im Breisgau, 1930–38), vol. 3, cols. 469–70; C. Spicq, *Esquisse d'une histoire de l'exégèse latine au Moyen Age* (Paris, 1944), 49–51.

    9. Christian Druthmar, *Expositio in Matthaeum evangelistam*, PL 106, 1261–1504, VI.1298D: . . . *permiserit ei sub una visione totum mundum ostendere, eo modo quo et beatum Benedictum legimus vidisse. Vel alio modo, species quaedam est, quae frequenter invenitur in Scriptura divina et vocatur Graece synechdoche, cum totum dicit, et pars intelligitur. Videlicet ostendit ei partem aliquam, et promisit ei reliquam cum toto daturum.*

    10. Gregory the Great, *Dialogues*, ed. A. de Vogüé, 2 vols. (Paris, 1979), II.35.3: *Mira autem ualde res in hac speculatione secuta est, quia, sicut post ipse narrauit, omnis etiam mundus, uelut sub uno solis radio collectus, ante oculos eius adductus est.*

Haimo of Auxerre, a Carolingian scholar active from 840 to 875, offers a similar explanation in his *Homilies* but adds an important touch.[11] He suggests that Christ, by his own will and power, not by those of the devil, could see all the world in two ways—either embracing it at once in a supernatural manner, although by means of human vision, or seeing it on a map or globe, in the same way as some saints had seen it.[12]

Although Haimo does not directly mention St. Benedict's vision, the resemblance between his text and Gregory the Great's invites this association. The words he chooses to describe the world as it appears to Christ—"as if collected on a globe [or map]" *[sive in sphaeram collectum]*—evoke Gregory's description of the world appearing to St. Benedict "as if collected under the ray of the sun" *[uelut sub uno solis radio collectus]*. Haimo further strengthens this parallel when he uses the same expression to describe the saints who were granted to see the world as if "collected on a map" *[in sphaeram collectum]*. The way Haimo puts together and compares the two possibilities of seeing the world—a contemplative vision and a map—suggests that he regards them as equivalent.

What do these interpretations reveal about early medieval perceptions of space and what are their implications for geographical knowledge? The only two ways to perceive the whole world, even for Christ, were a contemplative vision or a representation on a map or a globe. Amply documented in early medieval literature, contemplative cosmic visions embracing the earth at once and presenting it before the eyes of a spectator usually invited him and his listeners to contemplate the smallness, transience, and

---

11. On Haimo of Auxerre see R. Quadri, "Aimone di Auxerre alla luce dei 'Collectanea' di Heiric di Auxerre," *Italia medioevale e umanistica* 6 (1963): 1–48; D. Iogna-Prat, "L'oeuvre d'Haymon d'Auxerre: État de la question," in *L'école carolingienne d'Auxerre de Murethach à Remi 830–908*, ed. D. Iogna-Prat et al. (Paris, 1991), 157–79, esp. 157–58. His *Homilies* are edited in PL 118, under the name of Haimo of Halberstadt (bishop of that city in 841–53) On the *Homilies* and Haimo see H. Barré, *Les homéliaires carolingiens de l'école d'Auxerre* (Vatican, 1962); A. Matter, "Exegesis and Christian Education: The Carolingian Model," in *Schools of Thought in the Christian Tradition*, ed. P. Henry (Philadelphia, 1984), 90–105, esp. 97–98.

12. Haimo of Auxerre, *Homiliae de tempore,* PL 118, 11–746 199D: . . . *sed qui omnia creavit per Divinitatem, ipse omnia, juxta quod voluit, simul vidit per humanitatem: sive totum, ita ut est, sive in sphaeram collectum.* . . . *Nec mirum, si Dominus sic totum mundum prospicere potuit, qui etiam quibusdam sanctis hoc in munere praestitit, ut eum in sphaeram collectum videre possent* [. . . But he who created everything through divinity, by his will saw everything simultaneously through humanity: either the whole [world], as it is, or collected on a globe [or map]. And no wonder that, if the Lord could see the whole world, he also granted it to some saints so that they might see it [as if] collected on a sphere.] For the term *sphaera* meaning "map" see Gautier Dalché, *La "Descriptio mappe mundi,"* 89.

sinfulness of this world.[13] Patrick Gautier Dalché has demonstrated that medieval maps could function in the same way, as visual tools for meditation on all these questions.[14]

Likewise, in our commentaries, the world perceived at once, either in a vision or a map, represents mundane glory and invites the contemplation of its vanity. Hrabanus Maurus emphasizes precisely this aspect in his allegorical commentary on this passage, without discussing other questions of perception. He maintains that the moment of time in which Christ experiences his vision signifies "not so much the speed [of receiving the image] but rather the fragility of transitory power."[15]

To see the world as an object of contemplation, the spectator had to separate himself from it, to regard it from a distance, defined in spiritual rather than physical terms. The saints mentioned by the Carolingian scholars distance themselves from the world as they see it presented at once; Christ is about to reject the temptation of its glory. To show the world on a map required a somewhat similar operation—visualizing the world in one's mind, from a distance, and drawing the resulting image. In both cases, the image of the world, seen from a distance and embraced as a whole at once, in a single moment of time, represented a certain abstraction, a theoretical projection of the physical world in the mind.

The perception of the world as an object of contemplation is reflected in some geographical descriptions. In certain historical writings, geographical surveys, preparing the scene for the action of divine providence manifested in human events, invited the reader to begin by contemplating the world and its connections to people. Orosius, in his *Seven Books of History against the Pagans,* evokes the contemplative tradition of viewing the earth. In the geographical introduction, he describes the earth as if seen from a watchtower, from a distance that allows one to contemplate its sins and misfortunes.[16] Gildas, who wrote his *On the Fall and Conquest of*

---

13. On early medieval cosmic visions and their Neoplatonic roots see P. Courcelle, "La vision cosmique de saint Benoît," *Revue des études augustiniennes* 13 (1967): 97–117; for additional examples of saints' cosmic visions see Gautier Dalché, "De la glose," 754.

14. Gautier Dalché, "De la glose," esp. 753–57, where he compares the descriptions in cosmic visions and map images.

15. Hrabanus *Commentar. in Matthaeum* I.4.784C: *Quod enim in Luca scriptum est, ostensa sibi omnia regna orbis terrae in momento temporis (Luc. IV), non tam conspectus celeritas indicatur, quam caducae fragilitas potestatis exprimitur. In momento enim cuncta illa praetereunt, et saepe honor saeculi abiit, antequam venerit.*

16. Orosius *Hist.* I.1.14. Questions about this text will be addressed in detail in chap. 3; on Orosius and the tradition of contemplative observation see also Gautier Dalché, "De la glose," 754–55.

Britain in the mid–sixth century, also invited his readers to contemplation (although of a different theme), in his geographical survey of Britain. His description of Britain, which is presented as an antipode of the sinful world, with all the features of Paradise, begins his story of the barbarian invasion and the destruction of Roman civilization among the Britons, a story that Gildas interprets as typologically signifying the original sin, God's judgment, and the way to salvation.[17] The image of Britain in its prelapsarian state, with all the biblical associations behind it, may have served as a starting point for the contemplation of the crucial Christian issues of the Fall, sin, and salvation.

The mental image of the world, used for contemplation, was the product of intellectual rather than empirical activity. It could be received in a vision, or drawn from maps, or derived from books. No personal experience was required in order to achieve this image. However, a certain part of the world, because of its special significance, had to be perceived and studied in a different way. Another biblical episode reveals this second kind of perception.

"All the land which you see I will give to you. . . ."

In Genesis, God commands Abraham: "Lift up your eyes, and look from the place where you are, to the north and to Africa and to the east and to the sea; for all the land which you see I will give to you and to your descendants for ever. . . . Arise, walk through the length and the breadth of the land, for I will give it to you."[18]

In explaining this episode, some Latin exegetes addressed the question of how Abraham could see the whole land of Israel, using both allegorical and literal interpretations.[19] I shall focus on literal explanations, with one exception where the allegory sheds some light on our study question.

This is the case with Ambrose of Milan (ca. 339–97), who bases his allegorical explanations on certain geographical assumptions. First of all, he understands God's promise as including the whole world, with its four

---

17. Gildas, *De excidio et conquestu Britanniae,* ed. T. Mommsen, MGH AA XIII (Berlin, 1898; reprint, 1961), 1–85 (page numbers in subsequent citations of Gildas De *excidio* refer to this edition). On Gildas' work and its purpose see Hanning, *Vision of History,* esp. 45–46 and 50–62.

18. Gen. 13:14, Revised Standard Version.

19. As in the previous case, I have used the *Patrologia Latina* and *CETEDOC* databases.

parts—east, west, north, and south. As it was manifestly impossible for Abraham to walk around the world, Ambrose interprets "the whole world" as a symbol of spiritual rewards given the faithful.[20]

The order to walk around, then, must also be understood allegorically, as a spiritual journey directed away from corporeal things. "For," explains Ambrose, "he [Abraham] could not momentarily walk around the earth that included the empire of the Persians, from the shores of India to the columns of Hercules, as they are called, or to the extreme limits of Britannia. He would even show a lack of faith and obedience to the heavenly oracle, if he accepted the order [as meaning] to walk around this earth."[21]

Unlike Ambrose, Jerome (345–420) explains this passage literally. He limits God's promise to one country by saying that Abraham just had to look in the four cardinal directions—the north; the south, signified by the name of Africa; the east; and the west, designated by the sea, because "the region of Palestine is situated in such a way that it has the sea to the west."[22]

In Jerome's interpretation, this passage means the physical Palestine. To define its real boundaries, Jerome turns to the geographical tradition: "There is no doubt that Egypt is near Palestine, for God told Moses: I will give you all the land from the river of Egypt to the great river Euphrates, so that [Palestine] has on the one side the Egyptians, on the other the Assyrians. . . ."[23] Jerome does not comment on how Abraham could see Palestine.

---

20. Ambrose, *De Abraham,* ed. C. Schenkl, CSEL 32–1 (Vienna, 1897), 498–638, II.7.37: *oriens enim et occidens et septentrio et meridies portiones sunt uniuersitatis; his enim totus orbis includitur. haec cum promittit deus daturum se Abrahae, quid aliud declarat nisi sapienti et fideli praesto omnia, deesse nihil?*

21. Ibid., 595: *utique intra momentum terram istam Persarum interclusa imperiis, ab Indiae quoque litoribus usque ad Herculis ut aiunt columnas uel Brittanniae extrema confinia non potuit perambulare. et potuit quasi indeuotus uideri, qui caelesti oraculo non oboedisset, si obeundae huius terrae mandatum accepisset.*

22. Jerome, *Hebraicae quaestiones in libro Geneseos,* ed. P. de Lagarde, CCSL 72 (Turnhout, 1959), 1–56, 13.14.15, p. 17 (page numbers in subsequent citations of Jerome *Hebr. quaest. in Gen.* refer to this edition): <u>*Leua oculos tuos et uide a loco, in quo nunc es, ad aquilonem et ad austrum et ad orientem et ad mare: quia omnem terram, quam tu uides, tibi dabo eam et semini tuo.*</u> *quatuor climata mundi posuit, orientem et occidentem, septentrionem et meridianum. Quod autem in omnibus scripturis legitur, hic semel dixisse sufficiat mare semper pro occidente poni ab eo, quod Palaestinae regio ita sita sit, ut mare in occidentis plaga habeat.*

23. Jerome, *In Hiezechielem,* CCSL 75 (Turnhout, 1964), V.16.23/26, p. 188: *Nulli autem dubium quod Aegypto uicina sit Palaestina, dicente Domino ad Abraham:* <u>*Dabo tibi omnem terram, a fluuio Aegypti usque ad flumen magnum Euphraten,*</u> *ut ex una parte Aegyptios, ex altera habet Assyrios. . . .*

In comparison, Augustine's explanation of this passage focuses on the question, How could Abraham see all the Promised Land from the place where he was standing? Augustine answers by pointing out that Abraham could not and did not, in fact, see all of it from one place, because the Scripture mentions "the land which you see," not "the land as far as you see." He maintains that it was in order for Abraham to see all Palestine with his physical eyes that God ordered him to walk around it, "so that by walking around he might reach the land that he could not see with his eyes while staying in one place."[24]

Augustine emphasizes that Abraham had to see the Promised Land with his eyes rather than simply imagine it in his mind—in Augustine's own terms (which he explains elsewhere), by the corporeal rather than spiritual or intellectual vision. According to Augustine's idea of the three possible kinds of vision, the first, corporeal, perceives material things of the physical world through the senses of the body: "through it we see heaven and earth and everything in them that meets the eye." The second kind, spiritual, perceives the images of absent material things through the spirit. Augustine writes that by means of it, "we think of heaven and earth and the visible things in them even when we are in the dark." He adds, "In this case we see nothing with the eyes of the body but in the soul behold corporeal images." He further explains, "My manner of thinking about Carthage, which I know, is different from my manner of thinking about Alexandria, which I do not know." The third kind of vision, intellectual, perceives immaterial ideas, such as love, by means of intellect.[25]

---

24. Augustine, *Quaestionum in Heptateuchum libri VII*, ed. I. Fraipont, CCSL 33 (Turnhout, 1958), 1–377, *Quaest. Gen.* XXVIII, pp. 11–12 (page numbers in subsequent citations of Augustine *Quaest. Gen.* refer to this edition): *Quaeritur hic quomodo intellegatur tantum terrae promissum esse Abrahae et semini eius, quantum poterat oculis circumspicere per quattuor cardines mundi. Quantum est enim, quod ad terram conspiciendam acies corporalis uisus possit adtingere? . . . non enim dictum est: tantum terrae tibi dabo quantum uides, sed: tibi dabo terram quam uides. Cum enim et ulterior undique dabatur, profecto haec praecipue quae uidebatur dabatur. Deinde adtendendum est quod sequitur, quoniam, ne putaret etiam ipse Abraham hoc solum promitti terrae quod aspicere uel circumspicere posset,* <u>surge</u>*, inquit,* <u>et perambula terram in longitudine eius et latitudine, quia tibi dabo eam</u>*, ut perambulando perueniret ad eam, quam oculis uno loco stans uidere non posset.*

25. Augustine *De Gen. ad lit.* XII.6, p. 387 (trans. vol. 2, pp. 185–86): *tria visionum genera occurrunt: unum per oculos, quibus ipsae litterae uidentur, alterum per spiritum hominis, quo proximus et absens cogitatur, tertium per contuitum mentis, quo ipsa dilectio intellecta conspicitur. in his tribus generibus illud primum manifestum est omnibus; in hoc enim uidetur caelum et terra et omnia, quae in eis conspicua sunt oculis nostris. nec illud alterum, quo absentia corpora corporalia cogitantur, insinuare difficile est; ipsum quippe caelum et terram et ea, quae in eis uidere possumus, etiam in tenebris constituti cogitamus . . . aliter enim cogitamus Carthaginem, quam nouimus, aliter Alexandriam, quam non nouimus. tertium uero illud, quo dilectio intellecta conspicitur.* See also the commentary and bibliographical references in Augustine, *Literal Meaning*, 301–7.

In light of this idea, Augustine's insistence that Abraham had to perceive the Holy Land with his corporeal senses adds emphasis to the literal interpretation of Palestine as a concrete physical place. To fulfill God's command and reach that place by his eyes, Abraham had to physically walk around it.

Augustine's interpretation influenced several Carolingian scholars, who, without asking his question, picked up his emphasis on physical seeing and walking. Alcuin (d. 804) and Angelomus of Luxeuil (writing in the 840s) explain this in identical words: "Question: For if only the land that he could see then in four directions of the world, standing in one place, was promised him [Abraham], would the Promised Land seem to be small? Answer: Not only the land that he could see from this place was promised him. For it is not said: I shall give you as much land as you can see, but: I shall give you the land which you see. . . ." For this reason, they argue, God ordered Abraham to walk around the land, so that he might reach it by his eyes.[26] In their explanation of the four directions, both Alcuin and Angelomus follow Jerome.[27]

Remigius of Auxerre (ca. 841–ca. 908),[28] following Augustine's tradition and paraphrasing Alcuin and Angelomus, further emphasizes the necessity of walking around in order to examine the land: "It would not be much if he [God] gave [Abraham] all the land that he could see from the place where he stood. For that reason, it is added: rise and walk around the land. So that if you cannot examine its size with your carnal eyes, staying in one place, you could know how spacious it is by walking around it."[29]

---

26. Alcuin, *Interrogationes et responsiones in Genesin,* PL 100, 515–66, 169.535D–536A: *Si enim terra tantum promittebatur [ei], quam tunc in uno loco stans videre potuit per quattuor climata mundi; angusta videtur terra promissionis? - Resp. Non hoc [loco] solum terrae [ei] promissum est, quod videre potuit. Non enim dictum est, tantum terrae tibi dabo quantum vides; sed: Tibi dabo terram, quam vides, cum et ulterior undique dabatur. Et idcirco subjunxit: Surge et perambula terram in latitudine et longitudine [ejus] (vers. 17), ut perambulando perveniret ad ea, quae oculis uno loco stans videre non potuisset.* Angelomus of Luxeuil copies this text without using the question-and-answer format; see his *Commentarius in Genesin,* PL 115, 107–244, XIII.14.173A–B.

27. Alcuin *Interrogationes et responsiones in Genesin* 161.536A = Angelomus *Commentar. in Genesin* XIII.14.173A–B: *Solet [enim] Scriptura sancta, dum de plagis terrae promissionis loquitur, Mare pro Occidente ponere, ab eo, quod Palestinae regio Mare in Occidentis plaga habeat.*

28. On Remigius' Genesis commentary see B. V. N. Edwards, "The Two Commentaries on Genesis Attributed to Remigius of Auxerre, with a Critical Edition of Stegmüller 7195" (Ph.D. diss., University of Pennsylvania, 1992).

29. Remigius *Super Gen.* 2734–39: *Non multum erat si omnem terram illi daret quam cernere poterat de loco, in quo positus erat. Idcirco subiungitur: surge et perambula terram. Ut qui eius magnitudinem non potes uno in loco positus oculis carnis perpendere, perambulando illam noueris quam sit spatiosa.*

Other Carolingian scholars—Wigbod, who wrote his commentary on Genesis for Charlemagne between 775 and 800,[30] and Haimo of Auxerre[31]—who also understand this passage literally as meaning the real Palestine, do not have this emphasis. They only repeat Jerome's explanation of the four directions.[32]

The tradition started by Augustine strongly emphasizes that Abraham had to examine the Promised Land with his own eyes and that, in order to do that, he had to travel around it. As opposed to the whole world perceived by Christ in one moment of time, the Promised Land had to be examined in detail. Why is there such an emphasis, and what are its implications for the perception and description of the earth in geographical tradition?

The difference between the two episodes dictates the explanation closest at hand. Christ, being tempted by the vainglory of the world, rejects it; the world seen as an abstraction does not need examination. On the contrary, Abraham, by the covenant with God, receives Palestine for himself and his descendants. As his future possession, it has to be examined in detail.

The possession—particularly one aspect of it—attracted the attention of several exegetes. Palestine, promised to Abraham's descendants for eternity, had been apparently lost to the Jews: it was conquered by the Romans, and the Temple was destroyed. This contradiction required some explanation, provided by Augustine and Bede.[33]

---

30. On Wigbod see M. Gorman, "The Encyclopedic Commentary on Genesis Prepared for Charlemagne by Wigbod," *Recherches augustiniennes* 17 (1982): 173–201. Wigbod's commentary, partially edited in PL, is falsely attributed to Bede (PL 93, 233–430 and 96, 1101–68).

31. The commentary edited in PL under the name of Remigius of Auxerre has now been reattributed to Haimo (PL 131, 53–134.) On this matter see B. Edwards, "In Search of the Authentic Commentary on Genesis by Remigius of Auxerre," in *L'école carolingienne d'Auxerre de Murethach à Remi, 830–908,* ed. D. Iogna-Prat et al. (Paris, 1991), 399–412, and Edwards, "Two Commentaries."

32. Wigbod, *Quaestiones super Genesim,* PL 93, 303D: *Quatuor climata mundi posuit, orientem et occidentem, septentrionem et meridianum. Quod autem in omnibus Scripturis legitur, hic semel dixisse sufficiat, mare semper pro occidente poni, ab eo quod Palaestinae regio ita sit ut mare in occidentis plaga habeat.* Haimo of Auxerre, *Commentarius in Genesin,* PL 131, 51–134, XIII.14.83D: *Per haec quatuor climata terra promissionis ei promittitur. Quod autem in quibusdam locis mare ponitur pro Occidente, ideo fit quia terra repromissionis in mari Occidentali est.*

33. Here I only look at the explanations suggested in the commentaries on Gen. 13:14; to look at others would be far beyond the scope of this work. On this issue see F. Cardman, "Fourth-Century Jerusalem: Religious Geography and Christian Tradition," in *Schools of Thought in the Christian Tradition,* ed. P. Henry (Philadelphia, 1984), 49–64, esp. 49–50; E. D. Hunt, *Holy Land Pilgrimage in the Later Roman Empire, A.D. 312–460* (Oxford, 1982), 1–2.

Augustine, commenting on Genesis 13:14, asks how we can understand the words "eternal possession" there, if the people of Israel only possessed their land temporarily. He suggests that we treat the word "eternal" there as a figurative expression, used in the same sense in which Horace once used it, because neither the Scripture nor Horace could seriously talk about eternity with respect to temporal things, such as human life or human possessions.[34]

Although Bede does not consider other questions posed by Augustine, his interpretation of this passage concentrates on the words "eternal possession," further developing Augustine's idea. He compares the readings of various codices, one saying that God promised the land to Abraham "for eternity," another saying "for ages."[35] Bede prefers the second reading, arguing that the term should be so limited because no human possession in this life can be eternal. Indeed, the people of Israel have been expelled from Jerusalem. However, they still live and will live in other cities of the land of Canaan, sharing them with the Christians, who can also be considered the descendants of Abraham, in the spiritual sense.[36]

Bede also considers another possibility for explaining this passage, one that completely removes the difficulty. Allegorically, the Promised Land can be understood as the kingdom of heaven: "for all the elect, that is, the descendants of Abraham, perceive the kingdom of heaven, signified by the Promised Land, as the one where they would remain forever."[37]

Palestine, however, was not only significant as a possession. It was also the Promised Land, the land given by God. In addition, it was the land

---

34. Augustine *Quaest. Gen.* XXXI, p. 13: <u>Dabo tibi et semini tuo post te terram, in qua habitas, omnem terram cultam in possessionem aeternam.</u> Quaestio est quomodo dixerit <u>aeternam</u>, cum Israhelitis temporaliter data sit.... An potius locutionis est scripturarum?... Sicut ait Horatius: <u>seruiet aeternum qui paruo nesciet uti.</u> Non enim potest in aeternum seruire, cuius ipsa uita, qua seruit, aeterna esse non potest. Quod testimonium non adhiberem, nisi locutionis esset.... (Cf. Horace *Epistulae* I.10.41.)

35. Bede, *Libri quatuor in principium Genesis*, ed. C. W. Jones, CCSL 118A (Turnhout, 1967), III.13.14–15, p. 180 (page numbers in subsequent citations of Bede *In Gen.* refer to this edition): <u>Omnem terram quam conspicis tibi dabo et semini tuo usque in sempiternum.</u> Quidam codices habent "usque in seculum," quod utrumque ex uno greco, quod est αιων, transfertur.

36. Ibid.: *Si ergo legatur "usque in aeternum," merito mouet quomodo <u>in sempiternum</u> possidere ualeat semen Abrae terram illam, cum humana conuersatio in hac uita non possit esse "sempiterna." Si uero legatur "in seculum" sicque accipiatur, quemadmodum fideliter tenemus, initium futuri seculi a fine praesentis ordiri, nil quaestionis mouebit, quia etsi expulsi sunt Israelitae de Hierosolimis, manent tamen in aliis ciuitatibus terrae Chanaan, et usque in finem manebunt et uniuersa terra illa cum christianis inhabitatur, et ipsum semen est Abrahae.*

37. Ibid.: *...quia regio patriae celestis, quam terra illa repromissionis figurabat, ita ab electis omnibus, semine uidelicet Abrahae, percipitur ut in ea sine fine permaneant....*

that both participated in and witnessed the events described in the Bible, from the first settlement of Israel to the incarnation and death of Christ. Christianity perceived this physical land as holy, with special status, resembling no other physical place on earth.[38] All these ideas, although not mentioned directly in the commentaries on Genesis 13:14, may have contributed to the reasons why Augustine and those who followed him emphasized the necessity to examine this land in detail, to see and experience it.

Thus, one place, because of its special status, differs from the rest of the world. Unlike the whole world, which could be perceived intellectually, it requires a detailed examination based on experience. In the Holy Land, every location, every physical detail of landscape is important. This perception, reflecting the special concern of medieval people with holy places, permeates their descriptions.

Experiencing and Describing the Holy Land

The wish to experience the Holy Land—to walk through it, to see and probably touch the holy places mentioned in the Bible—was inspiring pilgrims throughout the Middle Ages.[39] Egeria, who traveled to Palestine in the late fourth or fifth century and left a very detailed account of her journey,[40] particularly emphasizes the importance of physically reaching and seeing the holy places. She is most grateful to God for allowing her, "unworthy and undeserving, to walk and visit all the places."[41] She painstakingly describes the way to this or that place: the distance she had to cover, the fact that she had to do it on foot, the hardships on the way to

---

38. On these concepts in Judaism and Christianity see W. M. Davies, *The Gospel and the Land: Early Christianity and Jewish Territorial Doctrine* (Berkeley, 1974); on "holy space" see also Cardman, "Fourth-Century Jerusalem," 50–51.

39. On the early pilgrims see Hunt, *Holy Land Pilgrimage*. On the variety of motives for pilgrimages in the Middle Ages see J. Sumption, *Pilgrimage: An Image of Medieval Religion* (Totowa, 1975). On the wish to experience the Holy Land see J. Wilkinson, *Jerusalem Pilgrims before the Crusades* (Warminster, 1977), 40–41; Wilkinson quotes Paulinus of Nola, writing in 409: "The principal motive which draws people to Jerusalem is the desire to see and touch the places where Christ was present in the body" (Paulinus of Nola *Epist.* 49.402).

40. Egeria, *Itinerarium Egeriae,* ed. A. Francescini and R. Weber, CCSL 175 (Turnhout, 1965), 29–90; on her dates see ibid., 30.

41. *Itinerarium Egeriae* V.12: *Et licet semper Deo in omnibus gratias agere debeam, non dicam in his tantis et talibus, quae circa me conferre dignatus est indignam et non merentem, ut perambularem omnia loca, quae mei meriti non erant. . . .* Cf. G. F. M. Vermeer, *Observations sur le vocabulaire du pèlerinage chez Egérie et chez Antonin de Plaisance* (Nijmegen, 1965), 41; Vermeer suggests that in Egeria's usage, *perambulare* often means "to visit."

a particular spot. The considerable labor sometimes required to get to the holy places becomes bearable because, being devoted to God as are prayers offered at each location, it benefits from God's assistance.[42] Egeria also strongly emphasizes the importance of seeing and identifying the holy places, almost every time mentioning how they have been pointed out or shown to her.[43]

Bernard, a Frankish monk who visited the Holy Land and wrote his account ca. 865, begins with his wish to see the holy places of Jerusalem.[44] Although less often than Egeria, he still mentions how the holy places were shown to him. Like Egeria, he builds his description around walking: the verbs of movement connect the passages devoted to particular places.[45]

The locations that witnessed biblical events invited the pilgrims to prayer and inspired meditation that led them deeper into the sense of the Scripture. Physical places, testifying to the literal meaning of the biblical text, helped the literal interpretation of the Bible. Thus, pilgrimage experiences directly contributed to exegesis.[46]

The wish to convey these firsthand experiences, testifying to the truth of the Scriptures and helping to understand them, inspired pilgrims' accounts. The experience here had an independent value—it did not mat-

---

42. Here are only a few examples: *Itinerarium Egeriae* II.1: *Vallis autem ipsa ingens est ualde . . . in longo milia passos forsitan sedecim, in lato autem quattuor milia esse appellabant. Ipsam ergo uallem nos trauersare habebamus, ut possimus montem ingredi;* II.3: *Et quoniam nobis ita erat iter, ut prius montem Dei ascenderemus, qui hinc paret, quia unde ueniebamus melior ascensus erat . . .* ; III.1–2: *Qui montes cum infinito labore ascenduntur. . . . Et sic cum grandi labore, quia pedibus me ascendere necesse erat . . . tamen ipse labor non sentiebatur—ex ea parte autem non sentiebatur labor, quia desiderium, quod habebam, iubente Deo uidebam compleri. . . .*

43. Leo Spitzer demonstrated the importance of the verbs *ostendere* and *monstrare* in Egeria's text, as meaning "identifying and pointing out the holy places"; see L. Spitzer, "The Epic Style of the Pilgrim Aetheria," in *Romanische Literaturstudien 1936–1956* (Tübingen, 1959), 871–912, esp. 888–90.

44. Bernard, *Itinerarium Bernardi, monachi franci*, in *Descriptiones terrae sanctae*, ed. T. Tobler (Leipzig, 1874; reprint, Hildesheim, 1974), pp. 85–99, 1: *In nomine Domini volentes videre loca sanctorum, quae sunt Hierosolimis . . .*

45. Ibid., 13: *Inde perreximus in montem Oliveti, in cujus declivio ostenditur locus orationis Domini ad patrem. In latere autem praedicti montis ostenditur locus, in quo pharisaei deduxerunt ad Dominum mulierem in adulterio deprehensam;* 15: *Inde transivimus ad Bethaniam . . .* ; 16: *Ab Jerusalem transivimus ad Bethlehem.*

46. See Hunt, *Holy Land Pilgrimage*, 83–106; J. F. Taylor, *Christians and the Holy Places: The Myth of Jewish-Christian Origins* (New York, 1993), ix. Thomas O'Loughlin has also demonstrated how Adamnan (or Adomnan), in his *De locis sanctis*, made his description of the holy places serve exegesis; see Thomas O'Loughlin, "The Exegetical Purpose of Adomnan's *De Locis Sanctis*," *Cambridge Medieval Celtic Studies* 24 (winter 1992): 37–53. I am grateful to one of my anonymous readers for drawing my attention to this point.

ter much if it was recorded by pilgrims themselves or by other people from their words or even compiled from the two. What mattered to those who could not undertake the actual journey was that they could still share its spiritual benefits.[47]

Bede points out precisely this when he praises Adamnan's book written about the pilgrimage of Arculf (ca. 683–84):[48] "he [Adamnan] quickly committed to writing everything which Arculf had seen in the holy places which seemed to be worthy of remembrance. From this he made a book, as I said, which is useful to many and especially to those who live very far from the places where the patriarchs and apostles dwelt, and only know about them what they have learned from books."[49] Inspired by this feeling, Bede himself wrote about holy places, although he had never visited them. Using Adamnan's book and other "old writings," he composed a treatise that, as he hoped, would be useful for prayer and meditation.[50]

The importance of knowing every feature of the Holy Land also inspired works that took the form of reference tools—lists of biblical locations, composed from books rather than pilgrims' accounts. Jerome's *On the Location and Names of Hebrew Places* was a translation of Eusebius' *Onomasticon*;[51] Bede's *The Names of Places* was compiled from Jerome and Josephus Flavius.[52] Both works, like Jerome's *Interpretation of Hebrew Names*, which included places,[53] emphasize etymologies and the

---

47. J. Richard, "Les rélations de pèlerinage et les motivations de leurs auteurs," in *Wallfahrt kennt keine Grenzen: Ausstellung im Bayerischen Nationalmuseum, München, 28. Juni bis 7. Oktober 1984*, ed. T. Raff (Munich and Zurich, 1984), 143–54.

48. Adamnan, *De locis sanctis*, ed. L. Bieler, CCSL 175 (Turnhout, 1965), 175–234; on the date see 177.

49. Bede, *Bede's Ecclesiastical History*, ed. B. Colgrave and R. A. B. Mynors (Oxford, 1969), V.15, p. 509 (hereafter cited as *EH;* page numbers in subsequent citations to *EH* refer to the translation given in this edition): . . . *quaeque ille se in locis sanctis memoratu digna uidisse testabatur, cuncta mox iste litteris mandare curauerit. Fecitque opus, ut dixi, multis utile et maxime illis, qui longius ab eis locis, in quibus patriarchae uel apostoli erant, secreti ea tantum de his, quae lectione didicerunt, norunt.*

50. Bede, *De locis sanctis*, ed. I. Fraipont, CCSL 175 (Turnhout, 1965), 247–80, XIX.5: *Ex qua [Adamnani scripta] nos aliqua decerpentes ueterumque litteris comparantes tibi legenda transmittimus, obsecrantes per omnia, ut praesentis saeculi laborem non otio lasciui torporis, sed lectionis orationisque studio tibi temperare satagas.*

51. Jerome, *De situ et nominibus locorum hebraicorum*, in *Onomastica sacra*, ed. P. de Lagarde (Göttingen, 1887), 119–90.

52. Bede, *Nomina locorum ex beati Hieronimi presbiteri et Flavi Iosephi collecta opusculis*, ed. D. Hurst, CCSL 119 (Turnhout, 1962), 273–87. Bede also composed, from various literary sources, a list of place-names used in the Acts of the Apostles: *Nomina regionum atque locorum de Actibus Apostolorum*, ed. M. L. W. Laistner, CCSL 121 (Turnhout, 1983), 167–78.

53. Jerome, *Liber interpretationis hebraicorum nominum*, ed P. de Lagarde, CCSL 72 (Turnhout, 1959), 57–161.

biblical significance of place-names rather than actual locations. Both Jerome and Bede describe places by interpreting their names; once again, places, now represented by names, become the matter for exegesis. In fact, these authors maintain that the knowledge of names helps one to understand the Bible as much as an actual trip to the Holy Land. Jerome points out this essential equivalence: "He who has seen Judaea with his own eyes, and who knows the sites of ancient cities and places and their names, whether the same or changed, will gaze more clearly upon Holy Scripture."[54]

Jerome's and Bede's books, reflecting concern with detailed knowledge of the Holy Land, fulfilled an important practical need. The Bible frequently mentions places, both in and beyond Palestine; and biblical characters often travel.[55] Literal interpretations of the Bible thus had to address geographical matters; the level of detail depended on the interests of a particular scholar.

If we look at early medieval commentaries on the Book of Genesis, Jerome and Bede, not surprisingly, demonstrated the greatest attention to geographical questions, creating examples for later scholars to follow.[56] The anonymous author of the *Life of Bede* even put geography first when he listed Bede's numerous interests.[57] That Bede even exceeded Jerome in his wish to explain biblical topography is particularly obvious when he builds on Jerome's foundation.

The following example will serve to demonstrate this. Jerome, commenting on Abraham's route from Egypt to the Promised Land and

---

54. Jerome, *Praefatio Hieronimi in librum Paralipomenon juxta LXX interpretes,* PL 29, 423A: . . . *sanctam Scripturam lucidius intuebitur, qui Judaeam oculis contemplatus est, et antiquarum urbium memorias, locorumque vel eadem vocabula, vel mutata cognoverit.* The translation is modified by me from Hunt, *Holy Land Pilgrimage,* 94.

55. On the importance of space in the Hebrew Bible and on its "spacial sensitivity" see R. Cohn, *The Shape of Sacred Space: Four Biblical Studies* (Chico, Calif., 1981), 1–3.

56. Geographical comments appear in Jerome *Hebr. quaest. in Gen.* as follows: on the settlement of Noah's sons, at 9.18–10.29, pp. 10–14; on the travels of Abraham, at 12.9–13.4, pp. 16–18; on the travel of Lot, at 19.30, p. 23; etc. Geographical comments appear in Bede *In Gen.* as follows: on the sons of Noah, at II.9.27, pp. 139–40 and III.10.1–III.10.30, pp. 142–51; on the travels of Abraham, at III.12.4–III.12.10, pp. 171–73; etc. Carolingian scholars, for instance, Hrabanus Maurus and Remigius of Auxerre, draw geographical details from Jerome and Bede for their commentaries on Genesis.

57. *Vita Bedae Venerabilis* 3, PL 90, 37C: *In quibus [libris] terrarum regionumque diversarum situs, naturas, qualitates, subtiliter, ac si cuncta ipse peragrasset, plerumque describit.* . . . This passage is cited in C. W. Jones, "Some Introductory Remarks on Bede's Commentary on Genesis," *Sacris Erudiri* 19 (1969–70): 115–98 (the citation is at 120 n. 14), reprinted in C. W. Jones, *Bede, the Schools and the Computus,* ed. W. M. Stevens (Aldershot, 1994), chap. 4.

addressing the Scripture's statement that Abraham ascended from Egypt to the south and came to the city of Bethel (lying in the south of Palestine), suggests that Abraham did not turn to the south of Egypt, where the desert lay, but rather headed south of Bethel. However, one needs to think awhile before arriving at this sense of Jerome's explanation.[58] Bede, using Jerome's comment as a starting point, takes great care to account for geographical features and makes it very explicit. He spells out that when Scripture says that Abraham went up from Egypt to the southern regions, it means the south of Palestine, not Egypt: "Abraham is said to go up from Egypt because the land of Egypt is undoubtedly related to lie below not only the land of Canaan but also other regions of the world, just as the region of the Scythians is said to be on the opposite side, higher than other parts of the earth."[59] Bede goes on to support this statement with another geographical argument: the rivers flow from Palestine to Egypt as from a higher to a lower place.[60]

All descriptions of the Holy Land—in pilgrim accounts, reference books, and exegetical works—reflect the fact that it was perceived differently than the rest of the world. It always required detailed attention, sometimes based on personal examination, as opposed to the world as a whole, which could be perceived in a general way. There were other places that medieval people singled out because of their special significance. In what follows, I shall look at how the interrelations between biblical exegesis and geographical tradition shaped the representation of one such place.

Paradise Revisited

Medieval descriptions of Paradise are very numerous; many have been listed and studied by historians.[61] For my purpose, which is to highlight the relations between exegesis and geographical tradition and to show

---

58. Jerome *Hebr. quaest. in Gen.* 13.1–4, pp. 16–17: . . . *sicut in hebraeo scriptum est abiit in itinere suo per austrum usque Bethel. Idcirco enim de Aegypto profectus est, ut non desertum ingrederetur, quod cum Aegypto reliquerat, sed ut per austrum, qui aquiloni contrarius est, ueniret ad domum dei, ubi fuerat tabernaculum eius in medio Bethel et Ai.*

59. Bede *In Gen.* III.13.1, p. 175: . . . *de Egypto ascendere memoratur Abram, quia nimirum terra Egypti non solum terrae Chanaan, sed et ceteris mundi regionibus iacere perhibetur inferior, sicut e contra Scytharum regio aliis terrarum partibus fertur eminere sublimior.* On Egypt being situated at a lower level than other regions cf. Ambrose, *Exameron*, ed. C. Schenkl, CSEL 32–1 (Vienna, 1897), III.3.11, p. 67.

60. Bede *In Gen.* III.13.1, pp. 175–76: *Quod ex fluminum cursibus in hanc aliunde defluentium, ex illa profluentium, facillime potest conici.*

61. See Graf, *La leggenda;* Patch, *Other World;* Grimm, *Paradisus coelestis;* L. I. Ringbom, *Paradisus terrestris: Myt, Bild och Verklighet* (Helsinki, 1958).

their importance for other genres, I will only focus on select Genesis commentaries and geographical descriptions.

The earth described in the Bible differed from that described in classical geographical writings; Paradise was one of the places that had to be incorporated into the picture of the world inherited from antiquity. According to Genesis 2:8–10, after the creation of man, God put him in a garden in Eden, located in the East, pleasantly abounding with trees, and watered by a river that, flowing out of the garden, divided and became four rivers, named the Phison, the Geon, the Tigris, and the Euphrates. The Bible, without directly saying it, implies that the Garden of Eden, or Paradise, was located on earth.

The question whether Paradise should be understood in the literal sense (as a place on earth) or in the allegorical sense was part of the wider discussion about the interpretation of the Scripture (and, often, its reconciliation with classical knowledge). Both the literal and the allegorical traditions, Greek and Latin, contributed to it. By the time Augustine wrote his literal commentary on Genesis, which influenced all later Latin exegesis, both interpretations were firmly established.[62]

Augustine

Augustine lists three positions taken by exegetes: Paradise can be understood literally, "in the corporeal sense"; allegorically, "in the spiritual sense"; or in both senses—sometimes literally and at other times allegorically. He himself joins the third position and accepts the notion of Paradise in a both literal and allegorical sense.[63] However, he strongly supports the primacy of the literal understanding of Paradise, as opposed to his own earlier figurative treatment. As he confesses, unable to achieve it then, now he wants to show how everything in Genesis "is to be understood first of all not in the figurative but in the proper sense. . . ."[64]

For Augustine, to accept Paradise in the literal sense, as a real "delight-

---

62. For a survey of pre-Augustinian tradition see J. Daniélou, "Terre et paradis chez les Pères de l'Eglise," *Eranos Jahrbuch* 22 (1953–54): 432–72; Grimm, *Paradisus coelestis*, 22–54; and, unfortunately not discriminating enough, Patch, *Other World*, 134–43. On Augustine's setting up a tradition of the Paradise interpretation see Grimm, *Paradisus coelestis*, 20.

63. Augustine *De Gen. ad lit.* VIII.1, p. 229 (trans. vol. 2, p. 32): . . . *tres tamen de hac re quasi generales sunt sententiae. una eorum, qui tantummodo corporaliter paradisum intellegi uolunt, alia eorum, qui spiritaliter tantum, tertia eorum, qui utroque modo paradisum accipiunt, alias corporaliter, alias autem spiritaliter.*

64. Ibid., VIII.2, p. 232 (trans. vol. 2, p. 35): . . . *ut non figurate sed proprie primitus cuncta intellegerentur.* . . . The earlier book is *De Genesi contra Manichaeos*, PL 34, 173–220.

ful place shaded with fruit-bearing trees, spacious, and irrigated by an abundant source of water," means to accept a miracle of God that defies human experience.[65] Augustine tightly connects this miracle to another one, the creation of the first man, and emphasizes the link between the two: just as the first man was created from the earth, "Paradise, where God placed the man, should be understood as nothing else but a certain place on earth, that is, a land, where an earthly man would live."[66]

Augustine also defends the literal understanding of the four rivers flowing from the one river of Paradise. Following an established tradition, he identifies the Geon as the Nile and the Phison as the Ganges, explaining that they have changed their ancient names, unlike the Tigris and the Euphrates, which have kept theirs. Thus, he argues, all four names must signify true rivers, because "they are well known in the lands through which they flow and are spoken of in nearly all the world."[67]

He maintains that even the fact that "the source of some of these rivers is known and of others is completely unknown" should not prevent us from taking the account literally. He notes that since the exact location of Paradise is unknown to us, we should, trusting the biblical text, assume that the river there divides into four rivers that flow a great distance under the earth before they spring forth "in other places which have been designated as their sources." Augustine supports his argument by appealing to common geographical knowledge: everybody is aware that this happens in the case of certain rivers flowing from their sources via underground passages, even though the underground passages cover a shorter distance.[68]

---

65. Augustine *De Gen. ad lit.* VIII.1, p. 231 (trans. vol. 2, p. 34): ... *locum scilicet amoenissimum, fructuosis nemoribus opacatum eundemque magnum et magno fonte fecundum.* ...

66. Ibid., VIII.1, p. 229 (trans. vol. 2, pp. 32–33, modified by me): ... *paradisus, in quo cum conlocauit deus, nihil aliud quam locus quidam intellegatur terrae scilicet, ubi habitaret homo terrenus.*

67. Ibid., VIII.7, pp. 240–41 (trans. vol. 2, p. 43): *de his fluminibus quid amplius satagam confirmare, quod uera sint flumina nec figurate dicta, quae non sint, quasi tantummodo aliquid nomina ipsa significent, cum et regionibus, per quas fluunt, notissima sint et omnibus fere gentibus diffamata?* The identification of the Geon and the Phison may ultimately go back to Josephus Flavius *Antiquitates* I.1.3 (see Grimm, *Paradisus coelestis,* 121).

68. Augustine *De Gen. ad lit.* VIII.7, pp. 241–42 (trans. vol. 2, p. 43–44): *An eo mouebimur, quod de his fluminibus dicitur aliorum esse fontes notos, aliorum autem prorsus incognitos et ideo non posse accipi ad litteram, quod ex uno paradisi flumine diuiduntur? cum potius credendum sit, quoniam locus ipse paradisi a cognitione hominum est remotissimus, inde quattuor aquarum partes diuidi, sicut fidelissima scriptura testatur, sed ea flumina, quorum fontes noti esse dicuntur, alicubi isse sub terras et post tractus prolixarum regionum locis aliis erupisse, ubi tamquam in suis fontibus nota esse perhibentur. nam hoc solere nonnullas aquas facere quis ignorat? sed ibi hoc scitur, ubi non diu sub terris currunt.*

Augustine used geographical details to prove the literal meaning of the biblical text and to interpret it.[69] In this interpretation, he developed and transmitted to later Latin scholars the tradition that placed Paradise on earth and firmly connected a unique and holy place, created by God's miracle and inaccessible to human experience, to common places on earth, well known from experience and literary tradition.

But we should not forget about another, spiritual dimension of the notion "Paradise" that Augustine developed in his allegorical interpretation, again building on the established tradition.[70] As an allegory, Paradise symbolized the life of the blessed and spiritual delights; its trees, wisdom and learning; its four rivers, the four virtues. It could also be understood as the church itself, with the four rivers symbolizing the four Gospels.[71]

Both interpretations, proclaimed equally important by Augustine, coexisted and complemented each other in his thought. The allegorical understanding added a spiritual dimension that will always loom behind the literal one. The allegory did not even have to be directly named; the whole range of spiritual meanings was there, ready at hand and constantly present in the mind of medieval scholars. The terrestrial Paradise could easily turn into the celestial one; the latter lent additional holiness to an already holy place.

Isidore

Thus established in biblical exegesis, the concept of the terrestrial Paradise was transferred to the context of a geographical description by Isidore of Seville.[72] In the *Etymologies,* he begins his account of the provinces of Asia from Paradise, a place in the East.[73] Earlier, describing rivers, Isidore begins with the Geon, called the Nile among the Egyptians, and proceeds to the Ganges-Phison, the Tigris, and the Euphrates. Naming Jerome and

---

69. Cf. Taylor in Augustine, *Literal Meaning,* 2:256 n. 43.

70. The allegorical interpretation of Paradise goes back to Philo of Alexandria and the Alexandrian school of exegesis based on allegory. In the Latin West, Ambrose, among others, joined the allegorical interpretation. On Philo see Grimm, *Paradisus coelestis,* 22–32.

71. Augustine *De Gen. contra Manichaeos* II.9–10, cols. 202–4. For a detailed description see Grimm, *Paradisus coelestis,* 57–60.

72. In his commentary on Genesis Isidore treats this question only allegorically. The literal commentary he might have written is no longer extant (see Grimm, *Paradisus coelestis,* 77).

73. Isidore *Etym.* XIV.3.2: *Paradisus est locus in orientis partibus constitutus....*

the Roman historian Sallust as his authorities, he places the sources of both the Tigris and the Euphrates in Armenia.[74]

Not only does Isidore place Paradise and its rivers on the earth, as Augustine had done, but by listing them among other provinces and rivers known from classical geographical books, he firmly connects biblical geography to that of antiquity, biblical sacred place to ordinary places on earth. A biblical notion of Paradise, with all its implications, provides the point of departure for Isidore's geographical description. The geographical space of antiquity gets recategorized in biblical terms.

Behind this terrestrial Paradise is a celestial one, for Isidore, like Augustine, acknowledged both literal and allegorical interpretations. He talked about the double meaning of Paradise, or two Paradises: "one is terrestrial, where the first man bodily lived, another celestial, where the souls of the blessed are transported in order that they might immediately depart from their bodies."[75]

Isidore's *Etymologies* adapted the exegetical model to the needs of geographical description. Very popular in the Middle Ages, this encyclopedia provided material used in many geographical and exegetical works.

An anonymous geographical poem written between 636 and 736 closely follows Isidore.[76] Its description of Asia, just as in the *Etymologies*, begins with Paradise: "First of all, it [Asia] has the delights of the gardens of Paradise."[77] In the middle of it grows the tree of life; there is no excessive heat or cold; a spring that gives water to this pleasant place flows divided into streams.[78]

---

74. On the Geon-Niles see ibid., XIII.21.7; on the Ganges-Phison, XIII.21.8; on the Tigris and Euphrates, XIII.21.9–10; on the source of the Tigris and the Euphrates, XIII.21.10: *Sallustius autem, auctor certissimus, asserit (Hist. 4, 77) Tigrim et Euphraten uno fonte manare in Armenia. . . . Ex quo Hieronimus (Sit. et nom. 202) animadvertit aliter de Paradisi fluminibus intellegendum.*

75. Isidore, *Differentiae*, PL 83, 9–98, II.12.32, 75A–B: *De duplici paradiso. Unus est* terrenus, *ubi primorum hominum corporaliter vita exstitit, alter* coelestis, *ubi animae beatorum, statim ut a corpore exeunt, transferentur. . . .*

76. *Versus de Asia et de universi mundi rota*, ed. F. Glorie, CCSL 175 (Turnhout, 1965), 433–54 (page numbers in subsequent citations of *Versus* refer to this edition); on its date see 435.

77. *Versus* 3, p. 441: *Habet primum paradisi ortorum dilicias*. Cf. Isidore *Etym.* XIV.3.2: *Paradisus est locus in orientis partibus constitutus. . . . Quod utrumque iunctum facit hortum deliciarum.* For detailed verbal parallels see the apparatus of Glorie's edition, pp. 441–42.

78. *Versus* 3–4, pp. 441–42: *Habet etiamque uite lignum inter medias. / Non est estus neque frigus, sincera temperies, / Fons manat inde perennis fluitque in riuolis.* Cf. Isidore *Etym.* XIV.3.2–3.

The Isidorean model was reflected in an eighth- or ninth-century school text composed in the form of questions and answers.[79] The student, answering the teacher's question about the provinces of Asia, begins the list with Paradise, following the *Etymologies:* "Which are its [Asia's] main provinces?—It has Paradise, India, Aracusia. . . ."[80]

Some of the maps based on Isidore's *Etymologies* include Paradise. So, in a Vatican manuscript of the eighth or ninth century, the map shows in the East "Paradise, the land of Eden," and four rivers—the Ganges-Phison, the Nile (without its biblical name of Geon), the Tigris, and the Euphrates.[81]

Bede

Both exegetical and geographical traditions contribute to Bede's treatment of Paradise. In his commentary on Genesis, Bede follows Augustine's literal interpretation of Paradise as a place on earth. Like Augustine, Bede emphasizes its inaccessibility to human experience and knowledge. Like Isidore, he locates it in the East. According to Bede, Paradise lies very far away from places where people now live, beyond the ocean or unknown lands. He writes, "Indeed, only God would know whether it is there or elsewhere; however, we cannot doubt that this place existed and continues to exist on earth."[82]

Bede identifies Paradise as a place of grace, where the first man lived in this state before committing the original sin. Now, he argues, as a consequence of this sin, Paradise is lost to people and even its location is unknown. However, he adds, God has provided for a certain resemblance

---

79. Gautier Dalché, who has discovered this text in a twelfth-century manuscript (Paris, BN lat. 4892, s. XII, fols. 245r–v), thinks that it goes back to the eighth or ninth century and reflects Spanish cultural influence. See Gautier Dalché, *La "Descriptio mappe mundi,"* 97, where he also mentions his forthcoming edition of this text.

80. Paris, BN lat. 4892, fol. 245r: *Cuius hec sunt principales prouinciae? Habet Paradisum, Indiam, Aracusiam. . . .*

81. Vatican, Vat. lat. 6018, s. VIII–IX. The map is reproduced in R. Uhden, "Die Weltkarte des Isidorus von Sevilla," *Mnemosyne* 3, no. 1 (1935): 28. The text is transcribed in Uhden, "Die Weltkarte," and in *Mappa Mundi e codice Vatic. Lat. 6018,* ed. F. Glorie in CCSL 175, pp. 457–66. I quote the latter, p. 457: *Paradisus, terra Eden; Gandis flumen, Fison.*

82. Bede *In Gen.* I.2.8, p. 46: *. . . in orientali parte orbis terrarum sit locus paradisi, quamuis longissimo interiacente spatio uel oceani uel terrarum a cunctis regionibus quas nunc humanum genus incolit secretum. . . . Verum siue ibi siue alibi sit, Deus nouerit; nos tamen locum hunc fuisse et esse terrenum dubitare non licet.* Cf. Grimm, *Paradisus coelestis,* 80–81.

of our earth to the land that the first man once possessed and to which people will eventually return as to their celestial motherland. He argues that Paradise thus serves as a reminder of the lost grace and as a goal to strive for and that the Nile, or the Geon, at the same time exemplifies the river of Paradise and connects it to the rest of the world. The Nile, he writes, just like the river irrigating Paradise, brings water to the fields of Egypt; it also directly flows from the paradisiacal river.[83]

According to Bede, the four rivers flowing from Paradise also link it to the rest of the earth: Phison, alias Ganges, begins in the mountains of the Caucasus; the Nile, in Mount Atlas, the ultimate western boundary of Africa; the Tigris and the Euphrates, in Armenia.[84] For this concrete information, Bede turns to geographical descriptions. He gives the source of the Nile and the location of Mount Atlas according to Orosius, the sources of the Tigris and Euphrates according to Isidore.[85] Describing the land of Euilath, the region of India, rich in gold, where the Ganges flows, he turns for support to Pliny the Elder (perhaps here consulted for the first time in the history of Genesis commentaries): "And Plinius Secundus narrates that the regions of India, above all others, abound with gold."[86]

Bede's interpretation of Paradise, drawing on both exegetical and geographical traditions, established one more model for later scholars. His text, with its attention to geographical details, was reproduced in the writings of Carolingian exegetes, such as Hrabanus Maurus and Remigius of Auxerre.

Hrabanus Maurus

Hrabanus Maurus, both in his Genesis commentary and in his encyclopedic work *On the Natures of Things,* joins the Augustinian tradition and

---

83. Bede *In Gen.* I.2.10, p. 48: . . . *quo in hac quam nos incolimus terra Nilus plana inrigat Egypti. . . . Et prouida utique dispositione Dominus ac conditor rerum in nostro orbe uoluit habere similitudinem nonnullam patriae illius ad quam possidendam in primo parente creati sumus, ut ad promerendum eius reditum de uicino nos admoneret exemplo—maxime et flumine illo quod de paradiso constat emanare, Nilus namque qui inrigat Egyptum, ipse est Geon qui in sequentibus de paradiso procedere memoratur.* Cf. Grimm, *Paradisus coelestis,* 81–82.

84. Bede *In Gen.* I.2.10–11, p. 48: *Constat astruentibus certissimis auctoribus horum omnium fluminum quae de paradiso exire referuntur in nostra terra fontes esse notos: Phisonis quidem, quem nunc Gangem appellant, in locis Caucasi montis; Nili uero, quem scriptura ut diximus Geon nuncupat, nunc procul ab Atlante monte, qui est ultimus finis Africae ad occidentem; porro Tigris et Euphrates ex Armenia...*

85. Orosius *Hist.* I.2.29 and I.2.11; Isidore *Etym.* XIII.21.9–10.

86. Bede *In Gen.* I.2.11–12, p. 49: *Et Plinius Secundus narrat Indiae regiones auri uenis prae ceteris abundare terris.* Cf. Pliny *Nat. hist.* VI.21.80. See Grimm, *Paradisus coelestis,* 80.

combines literal and allegorical interpretation. In the Genesis commentary, he follows Bede's treatment of Paradise, reproducing it word for word.[87] As a consequence, all Bede's geographical references and emphases are present in Hrabanus' literal explanation. We there find Paradise located far away, beyond the ocean, and concealed from human eyes; the Nile, in resemblance of the river flowing in Paradise, waters Egypt; the four rivers emerge from their hidden sources on the earth.

But Hrabanus adds an extensive allegorical interpretation borrowed from other sources.[88] In this interpretation, Paradise symbolizes the church; its river, Christ, "flowing from the source of the Father"; the four rivers, the four virtues, "because while they pour into the heart, it is restrained from the heat of all carnal desires." In Hrabanus' interpretation, the four rivers can also signify the four Gospels; the trees, the saints; the fruits, the deeds of the saints.[89]

In the geographical books of the *De rerum naturis,* Hrabanus uses the same approach, but he chooses as his model Isidore rather than Bede. As in the rest of his book, he borrows both general structure and contents from the *Etymologies* and consistently adds allegorical explanations.[90]

Hrabanus, even more than Isidore, emphasizes the distinction of Paradise from the Asian provinces. He puts it in a separate chapter, entitled "On Paradise," with the account of provinces in the following chapter, called "On Regions."[91] The literal content, however, he copies from Isidore word for word: Paradise is a place in the East, irrigated by one stream that flows out of it and divides into four rivers.[92] The rest of the chapter on Paradise (more than half of it) is devoted to various allegorical interpretations of Paradise and its four rivers, borrowed from exegetical

---

87. Hrabanus Maurus, *Commentariorum in Genesim libri quattuor,* PL 107, 439–670, I.12.476A–480C. On Hrabanus' sources see Grimm, *Paradisus coelestis,* 91–92.

88. Mainly Paterius' selections from Gregory the Great's *Moralia* (see Grimm, *Paradisus coelestis,* 92).

89. Hrabanus *Commentar. in Gen.* I.12.479C–D: *A principio autem plantatur paradisus, quia ecclesia catholica a Christo, qui est principium omnium, condita esse cognoscitur (Isid.) Fluvius de paradiso exiens, imaginem portat Christi de paterno fonte fluentis, qui irrigat Ecclesiam suam verbo praedicationis et dono baptismi. . . . (Greg.) Quatuor ergo paradisi flumina terram irrigant, quia dum his quatuor virtutibus cor infunditur, ab omni desideriorum carnalium aestu temperatur. Item allegorice quatuor paradisi flumina, quatuor sunt Evangelia ad praedicationem cunctis gentibus missa. Ligna fructifera, omnes sancti sunt; fructus eorum, opera eorum. . . .*

90. On Hrabanus' methods of work and his sources, in detail, see Heyse, *Hrabanus.*

91. Hrabanus *De rerum nat.* XII.3.334A–335A.

92. Ibid., XII.3.334A–B = Isidore *Etym.* XIV.3.2–4.

works by Isidore and Gregory the Great. These interpretations are essentially the same as in Hrabanus' commentary on Genesis.[93]

Following the path of Isidore, who had applied Augustine's exegetical concept of Paradise to his geographical description, Hrabanus fully transfers exegetical methods to his. As a result, in his picture, Paradise, like other geographic objects, very explicitly demonstrates how the two dimensions—literal and allegorical, material and spiritual—coexist. Paradise as a place belongs to the material earth, with its trees and rivers. Its allegorical significance, immediately reported, transfers it into a different, spiritual plane. The material geographical dimension always reveals another, spiritual one behind it. Like Isidore's, but even more so, Hrabanus' geography is structured by and centered on biblical and exegetical space.

Remigius of Auxerre

Remigius of Auxerre, in his commentary on Genesis, describing the location of Paradise and its four rivers, turns exclusively to the literal tradition of Bede and Augustine. Although he uses Hrabanus' Genesis commentary as well, he does not follow Hrabanus' allegorical interpretations.

Once again, we see Bede's description of Paradise as a beautiful place in the East, beyond the sea, far away from human habitation and knowledge.[94] Remigius insists on the literal understanding of terrestrial Paradise and turns to Augustine and Bede for arguments to support it. Augustine provides the general idea. Bede provides a concrete geographical example: just as in any place on the earth, the river of Paradise watered its land. Remigius quotes Genesis 2:10 and explains: "'And the river was flowing from the place of delight to irrigate Paradise.' It is clear that Paradise is not a spiritual place, as some think, but a certain terrestrial place, although most remote from all human knowledge. For how could it not be a place on earth, if the river flowing out of it irrigated its surface? . . . For it irrigated Paradise, that is, all beautiful and fruitful trees, just as today the Nile is said to irrigate the plain of Egypt."[95]

---

93. Hrabanus *De rerum nat.* XII.3.334A–335A. For the comprehensive list of sources see Heyse, *Hrabanus,* 114–15. Hrabanus treats the four rivers in the same way earlier, in book XI, turning to Isidore's text for the literal explanation, but adding allegorical treatment borrowed from elsewhere.

94. Remigius *Super Gen.* 959–65.

95. Ibid., 995–1004: *Et fluuius egrediebatur de loco uoluptatis ad irrigandum paradisum. (2,10) Hinc patet paradisum non ut quidam uolunt spiritalem, sed terrenum quendam esse locum, licet ab omni humana cognitione remotissimum. Quomodo enim terrenus locus non est, si fluuius inde egrediens superficiem illius irrigabat? . . . Irrigabat autem paradisum, id est omnia ligna pulchra et fructuosa, sicut et hodie usque fertur Nilus Aegypti planitiem irrigare.* Cf. Augustine *De Gen. ad lit.* VIII.7 and Bede *In Gen.* I.2.10, p. 48.

The image of a river bringing water to the fields is meant to evoke the image of Paradise. Here again we encounter Bede's idea about the resemblance between Paradise and the earth as an example and admonishment: "'And Abraham lifted up his eyes etc.' till 'it was all watered like the Paradise of God and like Egypt in the direction of Segor' (13.10). . . . For by his wise provision God wished that our earth resembled Paradise, which we, in our first parent, had been created to possess, so that this close example might admonish us to be worthy to return there."[96]

Remigius of Auxerre continued the tradition of literal interpretation of Paradise as a place existing on this physical earth and, even though inaccessible to people, linked to their world by physical rivers. Thus, the mere physical existence of Paradise, in the same geographical space as the rest of the earth, would serve the purpose of Christian contemplation and edification.

Paradise and the Earth—Identifications

The model developed by Augustine, Isidore, and Bede left some room for geographical speculation: it placed Paradise in the East without further specifying its location. Some exegetical and geographical works, influenced by other traditions going back to Greek exegesis, attempted to be more precise in this question. Some also suggested different identifications for two of the rivers flowing from Paradise, the Geon and the Phison. Some left the names unidentified, which could lead to curious geographical allusions.

The *Cosmography* of Anonymus Ravennatis (early eighth century) combines Isidorean tradition with that of Greek exegesis. The Ravenna cosmographer devotes considerable attention to the discussion of Paradise; it occupies almost five pages in Schnetz's edition.[97] Anonymus places Paradise in the farthest East, beyond India, "in the desert, impassable for people, in the oriental zone"; as his authority, he quotes a Greek exegete, Athanasius of Alexandria.[98] Anonymus maintains that Paradise

---

96. Remigius *Super Gen.* 2720–26: <u>Eleuatis itaque Abraham oculis et cetera usque quae omnia irrigabantur, sicut paradisus Domini, et sicut Aegyptus uenientibus in Segor. (13,10)</u> . . . *Prouida autem dispensatione Deus in nostro orbe uoluit haberi similitudinem paradisi, ad quem possidendum conditi in primo parente fueramus, ut ad promerendum eius reditum de uicino nos admoneret exemplo.* Cf. Bede *In Gen.* I.2.10, p. 48.

97. Anonymus Ravennatis *Cosmographia* I.6–10, pp. 5–10.

98. Ibid., I.6, pp. 5–6: . . . *trans Indorum Dimiricam-Evilath patriam in intransmeabilem ab hominum itinere eremum ad horientalem plagam ipse paradisus esse hostenditur, sicut hic testatur mihi sanctus Athanaxius Alexandriae episcopus.* . . .

is concealed from human eyes and that no traveler ever visited it; he argues that even Alexander the Great could not cross the desert beyond India. Furthermore, he maintains that no cosmographer ever described it.[99]

To support his localization of Paradise and to emphasize its remoteness, Anonymus refutes the theory, going back to antiquity, that two of the rivers flowing from it, the Tigris and the Euphrates, begin in the mountains of Armenia. According to him, Armenia cannot be the location of Paradise: it is not in the far east, its land is infertile because of cold weather and the mountains, and no sweet fragrance there would signal the Garden of Eden.[100]

Anonymus places the real sources of the Tigris and the Euphrates, as well as the sources of the Geon and the Phison, somewhere beyond India, whence they flow under the earth, only appearing on the surface near the Armenian mountains.[101] Speaking of the four rivers, Anonymus does not identify the Geon and the Phison, as if he did not feel it necessary to tie them to any earthly rivers. Only later in the text, when describing Egypt and mentioning the Nile, "the most extraordinary among rivers," does he add, almost in passing, "this Nile some people also have called the Geon."[102] Anonymus never identifies the Phison, as if leaving this to the discretion of the reader.

The cosmography of Pseudo-Aethicus, composed between the fifth and the eighth centuries,[103] also relies on the reader's biblical associations. It does not mention Paradise but points out one of its rivers. Speaking of the Nile and adding that it is also called the Geon, the cosmographer reports that its source is unknown: "it flows from more secret places but first appears in Ethiopia."[104] The unknown source and the name is all that connects the Nile in Pseudo-Aethicus' account to the biblical narrative. The

---

99. Ibid., I.6–8, pp. 6–7.

100. Ibid., I.8, pp. 7–8: *non ergo in Armeniorum terra ille suavissimus paradisus esse ostenditur. et quomodo in oriente paradisus esse ascribitur, si in Armenia esse putatur? . . . que Armeniorum terra infertilis esse ascribitur, immo magis prae omnibus orientalibus regionibus frigida esse clarificatur et plus montuosa magis quam plana asseritur.*

101. Ibid., I.8, p. 8: *et plerique gentiles phylosophi in suis expositionibus decreverunt de duobus fluminibus, quod de Transindorum patria, eremo, de loco investigabili per diversas procedunt patrias, quod ipsi Tigris et Euphrates invisibiliter discerpentes hinc inde terram per immensa miliariorum spatia iuxta Armenie montes manifestantur. . . .*

102. Ibid., III.2, p. 36: *Quam praefatam Egypti patriam rigat fluvius qui dicitur Nilus, qui est super diversa flumina praecipuus . . . quem Nilum alii Geon esse dixerunt.*

103. Pseudo-Aethicus, *Cosmographia*, in *Geographi latini minores*, ed. A. Riese (Heilbronn, 1878), 71–103 (page numbers in subsequent citations to Ps.-Aethicus *Cosmographia* refer to this edition); on the date see Nicolet and Gautier Dalché, "Les 'quatres sages,'" 193.

104. Ps.-Aethicus *Cosmographia* 45, p. 89: *. . . Nilus, qui et Geon appellatur, de secretioribus promitur, sed in exordio in Aethiopia videtur. . . .*

same is true for the Tigris—our cosmographer only mentions that its source is unknown because it flows under the earth until coming to the surface.[105]

The name *Geon* makes a second intriguing appearance in Pseudo-Aethicus' cosmography. Listing the rivers of Europe, the author mentions a river Geon beginning in the fields of Gaul.[106] Although the name must have evoked biblical associations then, just as it does now, the cosmographer gives no comment on this biblical-sounding name attached to a river in Europe.

The commentary on the Pentateuch from the Canterbury school of Theodore and Hadrian, composed between 650 and 750,[107] differs in its character from the dominant Latin exegetical tradition because it uses many Greek patristic sources generally unknown in the West. Both Theodore and Hadrian, sent to Britain by Pope Vitalian in 668, were learned in Greek and Latin exegesis and transmitted this learning to their school at Canterbury.[108]

Relying on the Greek patristic tradition, this commentary suggests a very precise location of the terrestrial Paradise, unusual for Latin exegesis.[109] The commentator reports two opinions: first, that Paradise was located in heaven; and second, that Paradise "was where the Holy City Jerusalem is now, since it is only twenty miles from where Adam is buried." In addition to the burial place of Adam, the commentator argues, the place where Cain killed Abel, that is, the city of Emmaus, is also not far from Jerusalem. All these facts of biblical geography, according to the commentator, prove his point.[110]

---

105. Ibid., 10, p. 76: *fluvius subtus exit latenter, et ob hoc ortus eius non conprehenditur, quoniam de obscuritate promitur....*

106. Ibid., 23, p. 82: *Fluvius Geon nascitur in Galliarum campis. influit in oceano occidentali. currit milia CCCCII.*

107. Ed. and trans. in B. Bischoff and M. Lapidge, *Biblical Commentaries from the Canterbury School of Theodore and Hadrian* (Cambridge, 1994), 298–385 (hereafter cited as *PentI*; page numbers in subsequent citations refer to this translation); on the time of composition see ibid., 1.

108. On their mission and activities in Britain see Bede *EH* IV.1–2, V.20, V.23. See also J. M. Wallace-Hadrill, *Bede's Ecclesiastical History of the English People: A Historical Commentary* (Oxford, 1988), 329–33; Bede, *Venerabilis Baedae opera historica*, ed. C. Plummer, 2 vols. in 1 (1896; reprint, Oxford, 1966), notes at 2:202–7; Bischoff and Lapidge, *Biblical Commentaries*, 1ff.

109. See Bischoff and Lapidge, *Biblical Commentaries,* 441.

110. *PentI.* 35, pp. 309, 311: *Multi dicunt paradisum fuisse creatum et supra aplanem collocatum, ibique hominem constitutum antequam praeuaricaret imperium Dei. Alii autem a principio creationis caeli et terrae paradisum credunt esse creatum ubi modo est sancta ciuitas, quia .xx. milia tantum distat ab ea ubi sepultus est Adam.*

Thus precisely located in biblical space, which, in its turn, is represented in the terrestrial geography of the Holy Land, Paradise is also linked by its rivers to the rest of the earth, described in secular tradition. Of the four rivers, this commentator only mentions the Phison, unexpectedly suggesting that it is "the same river as the Rhône, which in turn is the same as the Danube."[111] This identification, reflecting the commentator's vague ideas about the two rivers' sources, and possibly a conflation of the Greek exegetical tradition and classical geography, may have been the commentator's own contribution.[112]

The origin of this identification, however intriguing, is not the most interesting thing about it. Far more important for our purpose is the fact that the Canterbury commentator, by locating one of the Paradise rivers in Europe, explicitly connects this part of the world to the sacred place of the biblical narrative, with all its spiritual overtones and allegorical associations. Although based on Greek precedents, this connection is the first one in Latin exegesis known to me. It also seems to be the only one in the Canterbury commentaries.[113]

Another commentary on Genesis belonging to the same school does not mention the rivers at all.[114] It refers only to Paradise and, in doing so, combines Greek and Latin traditions. Following Isidore's *Etymologies,* the commentator places Paradise in the East and describes its trees, its temperate climate, and the fountain that divides into four rivers. By the end of the Isidorean passage, the commentator adds the Greek version: "Some say that Paradise is in the center of the earth where Jerusalem is, others think it was borne aloft in the air after man's sin; certain others suppose that it is located in the eastern sea."[115]

Attempting to locate Paradise precisely and identify the rivers flowing

---

111. Ibid., 37, *Phison . . . : i. eadem et Rodanus, ipse et Danubius.*

112. See the explanation suggested in Bischoff and Lapidge, *Biblical Commentaries,* 443 n. 37: some Greek exegetes identified the Phison with the Danube but not with the Rhône; some ancient geographers appear to confuse the sources of the Danube and the Rhône; the Canterbury commentator may have meant not the Rhône (Rhodanus) but the river of eastern Europe, the Duna (Rhoudanos).

113. As edited by Bischoff and Lapidge in *Biblical Commentaries.*

114. *Supplementary Commentary on Genesis, Exodus, and the Gospels,* ed. and trans. in Bischoff and Lapidge, *Biblical Commentaries,* 386–423 (hereafter cited as *Gn-Ex-Evla;* page numbers in subsequent citations refer to this edition).

115. *Gn-Ex-Evla* 9, 89: *De paradiso terrestri. Paradisus est locus in orientis partibus constitutus. . . . Est enim omni genere ligni pomiferarum arborum consitus, habens etiam lignum uitae. Non est ibi frigus, non aestas, sed perpetua aeris temperies. E cuius medio fons prorumpens totum nemus irrigat; diuiditurque in quattuor nascentia flumina. . . . Alii dicunt paradisum esse in medio terrae ubi Hierusalem est, alii eum putant post peccatum in aere leuatum; quidam in orientali mare collocatum esse uolunt.* Cf. Isidore *Etym.* XIV.3.2–4.

from it, early medieval scholars strove to establish a structure of geographical space that would carry spiritual meaning for them and their audience. The existence of Paradise as a physical place, located among other physical places and linked to them by well-known rivers, both made evident the literal truth of the Bible and invited one to contemplate its spiritual meaning. At the same time, the mere existence of Paradise in the same geographical space as the rest of the earth ennobled the latter and gave its people perspective and a spiritual goal to strive for.

Paradise and the Earth—Associations

The exegetical view of Paradise, with its several levels of meaning, contributed to the creation of the medieval picture of the world in geographical tradition. It also provided the background context—a certain set of associations ready in the mind of the medieval audience—on which an author could safely rely.

Certain geographical features that may have evoked the image of Paradise in the reader's mind—the beauty and fertility of the place, its pleasant and temperate climate, a great river—have, incidentally, been implanted in literature since antiquity. The topos of the *locus amoenus,* a pleasant place that had all the aforementioned characteristics, was familiar to the early Middle Ages from classical poets, Virgil most of all. Medieval writers often put it to use in their descriptions of Paradise.[116] Another ancient tradition mentioning the same features and inherited by medieval literature was that of praise of cities and countries.[117]

All these traditions worked together to provide the medieval set of associations. However, the image of Paradise, with all its meanings, seems to predominate in the works of hagiography and history that I shall now discuss. In them, the image of Paradise and the descriptions in which it was used served the authors' main purpose.

When the author of the *Deeds of the Holy Fathers of Fontenelle* (composed between 833 and 840)[118] described the location of his abbey, he cre-

---

116. On the *locus amoenus* in classical and medieval literature see E. R. Curtius, *European Literature and the Latin Middle Ages,* trans. W. R. Trask (1953; reprint, New York, 1963), 195–200.

117. On these eulogies in medieval literature, with numerous examples, see Curtius, *European Literature,* 157–59.

118. *Gesta sanctorum patrum Fontanellensis coenobii,* ed. F. Lohier and R. P. J. Laporte (Rouen and Paris, 1936); page numbers in subsequent citations of *Gesta* refer to this edition. On the date see P. Grierson, "Abbot Fulco and the Date of the *Gesta abbatum Fontanellensium,*" *English Historical Review* 55 (1940): 275–84.

ated an appropriate setting for the actions of the saints. Among other associations, it was meant to evoke the image of Paradise.[119] It is surrounded by fertile mountains and rich forests, and three rivers bring water to its pleasant meadows.[120] Moreover, one of the two rivers there bears the name of a river flowing from Paradise: "To the south [of the abbey is located] the largest of the rivers, the Geon, which is also [called] the Seine."[121] A little later, to support this biblical allusion, the author compares this great river to the Nile, a river traditionally identified with the Geon and serving, in exegetical tradition, as a special link with Paradise.[122]

The beauty of this allusion is in its complete reliance on the reader. The author never explicitly connects the Seine with Paradise; nor does he mention the traditional biblical identification of the Nile. He might have initially got his idea from the *Cosmography* of Pseudo-Aethicus, which names the Geon as a river in Gaul,[123] but he does not mention that either.

It was not even necessary to use biblical geographical names—only helped by allusions, the mind trained in exegesis would still search for other meanings behind the literal one and distinguish Paradise behind an earthly place. So Gildas, without directly referring to the Bible, very appropriately begins his story of the fall and conquest of Britain with a picture resembling Paradise. The Britain Gildas describes has distinctive paradisiacal features—noble rivers that give water to its pleasantly located valleys and hills, fields most appropriate for cultivating, mountains convenient for pasturing and decorated with beautiful flowers.[124] This image of

---

119. Jacques Fontaine has pointed out the spiritual symbolism of this passage, going back to the classical tradition of *locus asceticus* and *locus amoenus,* meant to evoke the experience of earlier saints turning desert into paradise. See J. Fontaine, "La culture carolingienne dans les abbayes normandes: L'exemple de Saint-Wandrille," in *Aspects du monachisme en Normandie, IVe–XVIIIe siècles. Actes du colloque scientifique de l' "Année des abbayes normandes"* (Paris, 1962), 31–54; on the symbolic landscape see 44–45.

120. *Gesta* 5, pp. 6–7: *A tribus enim plagis, id est a septentrionali, occidua atque australi, montibus arduis ac frugiferis bachique fertilissimis siluisque est obsitum condensis. Ab oriente item habet fontem huberrimum. . . . Inter haec duo mirabilia flumina prata eiusdem coenobii sunt amoena atque irrigua.*

121. Ibid. 5, p. 7: *Ab Austro item maximus fluuiorum Geon, qui et Sequana. . . .*

122. Ibid.: *Talique impetu per meatus praedictorum duorum fluminum perque prata illis contigua, ceu Nilus aegiptiacus per spatia passuum plus minusque D CCC ad murum eiusdem accedunt coenobii. . . .*

123. Cf. Ps.-Aethicus *Cosmographia* I.23, p. 82. Gautier Dalché sees this passage in the *Gesta* as a proof that the *Cosmography* was known at Fontenelle (Nicolet and Gautier Dalché, "Les 'quatres sages,'" 199).

124. Gildas *De excidio* 3, p. 28: *vallata duorum ostiis nobilium fluvium Tamesis ac Sabrinae veluti brachiis, . . . aliorumque minorum meliorata, . . . campis late pansis collibusque amoeno*

the prelapsarian Britain, with a picture of Paradise behind it that gave it another, spiritual dimension, serves Gildas as a starting point for his story of the vicissitudes of the Britons' fate, behind which the reader also had to distinguish another, spiritual dimension—the story of sin and salvation.

Isidore of Seville also uses allusions to Paradise when he begins his *History of the Goths* with the "Praise of Spain" and describes the land that belongs to the Goths, God's new chosen people.[125] He writes that this land is rich and abundant in fruits and grain; that it is situated in the most pleasant region of the world, neither hot nor cold but temperate; that it is watered by full rivers and golden streams. He adds that it produces great leaders and is now a fit home for "the most flourishing nation of the Goths."[126]

Bede begins his *Ecclesiastical History* with a description of Britain, which goes back to Gildas in its idea and in some particular details.[127] Bede's Britain and Ireland resemble both Paradise and the Promised Land: Britain is rich in crops and trees, abundant in rivers and hot springs, and produces metals; Ireland has a mild and healthy climate, it abounds in milk and honey, and its very air kills snakes.[128] As in Gildas' account,

---

situ locatis, praepollenti culturae aptis, montibus alternandis animalium pastibus maxime convenientibus, quorum diversorum colorum flores humanis gressibus pulsati non indecentem ceu picturam eisdem imprimebant. . . .

125. On the purpose of Isidore's *History* and on the Goths as God's new chosen race see J. H. Hillgarth, "Historiography in Visigothic Spain," in *La storiografia altomedievale,* Settimane di Centro Italiano di studi sull'alto medioevo, 17 (Spoleto, 1970), 261–311, esp. 298–99. On Isidore's *De Laude Spaniae* and its classical models see Curtius, *European Literature,* 157–58.

126. Isidore, *Historia Gothorum,* ed. T. Mommsen, MGH AA XI (Berlin, 1894. Reprint, 1961), 241–95, *De laude Spaniae* (trans., *Isidore of Seville's History of the Goths, Vandals, and Suevi,* trans. G. Donini and B. Ford [Leiden, 1970], 1–2): *tu bacis opima, uvis proflua, messibus laeta . . . tu sub mundi plaga gratissima sita nec aestivo solis ardore torreris nec glaciali rigore tabescis, sed temperata caeli zona praecincta zephyris felicibus enutriris. . . . tu superfusis fecunda fluminibus, tu aurifluis fulva torrentibus . . . opulenta es principibus ornandis ut beata pariendis. . . . denuo tamen Gothorum florentissima gens post multiplices in orbe victorias certatim rapuit et amavit, fruiturque hactenus inter regias infulas et opes largas imperii felicitate secura.*

127. Bede *EH* I.1. Cf. Gildas *De excidio* 3, p. 28.

128. The text on England (Bede *EH* I.1) reads: *Opima frugibus atque arboribus insula, et alendis apta pecoribus ac iumentis, uineas etiam quibusdam in locis germinans. . . . Habet fontes salinarum, habet et fontes calidos, et ex eis fluuios balnearum calidarum omni aetati et sexui per distincta loca iuxta suum cuique modum accommodos.* That on Ireland (ibid., I.1) reads: *Hibernia autem et latitudine sui status et salubritate ac serenitate aerum multum Brittaniae praestat . . . nemo propter hiemem aut faena secet aestate aut stabula fabricet iumentis; nullum ibi reptile uideri soleat, nullus uiuere serpens ualeat. . . .* On Bede's imagery of Paradise and the Promised Land see Kendall, "Imitation," 178–82, and Wallace-Hadrill, *Bede's Ecclesiastical History,* 6, 9.

Bede's "creation scene," imitating the Book of Genesis, evokes the image of the prelapsarian state of man: blessed by God in its nature, this land is about to receive the blessing of conversion and Christianity. Thus the image of Paradise begins Bede's story of the English people and their way from fall to salvation, the story of the English church and its progress.

The images of Paradise that must have been easily evoked in the mind of the medieval reader served to indicate a spiritual dimension behind descriptions of earthly places. They involved the reader in the contemplation of spiritual matters, such as the divine providence manifested in the deeds of saints or histories of nations. Medieval scholars could easily achieve this goal using the methods of allegorical and typological interpretation, perfected in biblical exegesis and transferred to the realm of geographical descriptions.

Exegetical writings reveal certain ideas about geographical space. The only way for early medieval people to perceive the whole earth or a whole country at once was in their mind.[129] This mental image, whether received in a vision, drawn on a map, or represented in verbal descriptions, was the product of intellectual rather than empirical activity. Early medieval scholars used this image to contemplate spiritual matters lying beyond the sphere of the physical world.

The contemplative goal was also pursued in the descriptions of the only place that required personal experience and attention to details—the Holy Land. Pilgrims' accounts and exegetical writings in their descriptions of Palestine aimed at helping the reader to approach the truth of the Bible. The knowledge of holy places mentioned in the Scripture was perceived as one of the ways to gain the knowledge of the Scripture. Descriptions of such places approached exegesis in their goal and were guided by exegetical methods.

Not only methods but paradigms of description born in exegesis influenced geographical knowledge. The descriptions of Paradise allow us to trace how this fundamental component of the medieval image of the earth, created in exegesis, became part of geographical descriptions that, in their turn, provided material for further exegetical works.

The image of Paradise, understood literally and allegorically, possessed both a material and a spiritual dimension. References or allusions to it in

---

129. See similar ideas about the "conceptual" or "imaginative" geography of antiquity in J. S. Romm, *The Edges of the Earth in Ancient Thought: Geography, Exploration, and Fiction* (Princeton, 1992), 9.

geographical or historical works evoked both perspectives in the mind of medieval scholars and invited those scholars to contemplate spiritual questions. Exegetical methods of interpretation facilitated this process.

The connections between biblical exegesis and geographical tradition functioned on many levels. Paradigms of description established in exegesis influenced geographical writings. Important components of the world image in the early Middle Ages go back to exegesis. Exegetical methods, transferred to certain geographical descriptions, both shaped and helped to interpret them. Finally, the goal of exegesis—to help people interpret the word of God—included the description and interpretation of places.

CHAPTER 3

History and Geographical Tradition

Historical and geographical questions were always perceived as related in early medieval thought. The present chapter will explore how the practice of writing history reflected this theoretical connection. To limit what could become a very broad research enterprise in its own right, I have chosen from all early medieval histories only those that begin with a geographical introduction, because the very fact that an author would decide to begin his history with a separate geographical section represents a certain statement about the relations between history and geography. I have further narrowed my search to four works, ranging from the early fifth to the tenth century.[1] Thematically, one addresses universal history, and three address the histories of single nations. I will approach each of them with the same questions: What was the goal of the geographical introduction and how did the author justify it? What did the introduction describe and

---

1. The approach I suggest is but one of many possibilities in the study of the role of geographical information and perceptions of space in early medieval historiography. Other authors need to be considered, for instance, Paul the Deacon who began his *Historia Langobardorum* with a description of *Germania* and *Scandinavia* (Paul the Deacon, *Historia Langobardorum*, ed. L. Bethmann and G. Waitz, MGH SRL [Hanover, 1878], I.1–2). Authors who, like Isidore of Seville, begin their histories with praises of places also deserve attention (Isidore *Historia Gothorum, De laude Spaniae*). Furthermore, the lack of a geographical introduction does not necessarily mean that an author demonstrates less attention to geographical space. Thus, Gregory of Tours' *Historiae* would be no less interesting to study than the four histories that I have chosen. The extent and significance of perceptions of geographical space in medieval historiography deserve a special study; promising directions have been explored in, for example, Witzel, *Der geographische Exkurs;* Brincken, "Mappa mundi und Chronographia"; M. Sot, "Organization de l'espace et historiographie épiscopale dans quelques cités de la Gaule caroligienne," in *Le métier d'historien au Moyen Age: Etudes sur l'historiographie médiévale,* ed. B. Guenée (Paris, 1977), 31–43; P. Gautier Dalché, "L'espace de l'histoire: Le rôle de la géographie dans les chroniques universelles," in *L'Historiographie médiévale en Europe: Actes du colloque, Paris, 29 mars–1er avril 1989,* ed. J.-Ph. Genet (Paris, 1991), 287–300.

how did it differ from geographical information given elsewhere in the same book? What tradition did it follow? And, finally, how did geographical descriptions serve the overall objective of the work?

The four works I will focus on have received unequal attention from scholars. Orosius' geography and its connection to history have been much studied.[2] Jordanes' geographical descriptions have attracted the attention of many scholars, who have mainly focused on the places of the Goths' origin and migration and, along the way, raised questions about Jordanes' geography in general.[3] Bede's concept of history has attracted more interest than his geography, but important work has been done on his introduction.[4] Richer's history has been far less studied.[5] Even though previous scholarship has focused on various aspects of the four authors' geographical and historical concepts, the aforementioned study questions both necessitate and justify another visit to this sometimes well-studied ground, eventually contributing to our understanding of how historians in the early Middle Ages perceived and practiced the connection between geographical and historical knowledge.

Orosius

Paulus Orosius, as he himself acknowledged, received the inspiration and the idea for his *Seven Books of History against the Pagans* from Augustine.

---

2. On Orosius' geography see K. Zangemeister, "Die Chorographie des Orosius," in *Commentationes philologae in honorem Th. Mommseni* (Berlin, 1877), 715–38; Y. Janvier, *La géographie d'Orose* (Paris, 1982). On the connections of geography and history see E. Corsini, *Introduzione alle "Storie" di Orosio* (Turin, 1968), the chapter entitled "La nuova geografia"; F. Fabrini, *Paolo Orosio, uno storico* (Rome, 1974), 312–30.

3. L. Weibull, "Scandza und ihre Völker in der Darstellung des Jordanes," *Arkiv för Nordisk Filologi* 41 (1925): 213–46; J. Svennung, *Jordanes und Scandia: Kritischexegetische Studien* (Stockholm, 1967); N. Wagner, *Getica: Untersuchungen zum Leben des Jordanes und zur frühen Geschichte der Goten* (Berlin, 1967), chap. 3; G. Dagron, "Discours utopique et récit des origines," pt. 1, "Une lecture de Cassiodore-Jordanès: Les Goths de Scandza à Ravenne," *Annales: Économies, sociétes, civilizations* 26 (1971): 290–305; W. Goffart, *The Narrators of Barbarian History (A.D. 550–800): Jordanes, Gregory of Tours, Bede, and Paul the Deacon* (Princeton, 1988), 88–96.

4. On Bede's introduction see Kendall, "Imitation." On Bede's concept of history see Hanning, *Vision of History,* 63–90; P. Hunter Blair, *Bede's Ecclesiastical History of the English Nation and Its Importance Today,* Jarrow Lecture, 1959 (Newcastle-upon-Tyne, 1959); Ray, "Bede, the Exegete."

5. There is no special work on Richer's geography, and there are very few monographs on his book in general; see the bibliography in H.-H. Körtum, *Richer von Saint-Remi: Studien zu einem Geschichtsschreiber des 10. Jahrhunderts* (Stuttgart, 1985). However, the editions of Richer's work and studies of tenth-century political history have addressed many geographical problems.

Orosius begins his book by mentioning Augustine's request to refute the pagans' claim that present times are beset by evils because the belief in Christ has replaced the belief in old gods. Orosius turns to past history to demonstrate that it contained much more misery, shame, and disaster than the present.[6]

He declares that a universal history, from the creation of the world to his own time, will serve his purpose: "I intend to speak of the period from the founding of the world to the founding of the City, then up to the principate of Caesar and the birth of Christ, from which time the control of the world has remained under the power of the City, down even to our own time."[7] In this declaration, Orosius demonstrates his view of the deep connection between historical events and the places where they occurred. Skillfully using the old pun on the words *urbis* and *orbis,* which goes back to Roman literature,[8] he ties the story of the world and the story of man (which goes *ab orbe condito usque ad Urbem conditam*) to the story of the physical earth and the story of the human institution most important to Orosius—the Roman Empire. In Orosius' concept of history, succinctly expressed in his statement of intent and developed later in his book, the earth or the world is destined to be controlled by the City *(sub potestate Urbis orbis mansit imperium)* and to accept Christianity.

Thus connected, the earth and the human race share the same fate, which Orosius wishes to demonstrate: "Insofar as I shall be able to recall them, I think it necessary to disclose the conflicts of the human race and

---

6. Orosius *Hist.* I, *Prologus* 1: *Praeceptis tuis parui, beatissime pater Augustine;* ibid., 9–10: *Praeceperas mihi, uti aduersus uaniloquam prauitatem eorum, qui alieni a ciuitate Dei . . . praesentia tamen tempora ueluti malis extra solitum infestatissima ob hoc solum quod creditur Christus et colitur Deus, idola autem minus coluntur, infamant:—praeceperas ergo, ut ex omnibus qui haberi ad praesens possunt historiarum atque annalium fastis, quaecumque aut bellis grauia aut corrupta morbis . . . per transacta retro saecula repperissem, ordinato breuiter uoluminis textu explicarem.*

7. Ibid., I.1.14 (trans. p. 7): *Dicturus igitur ab orbe condito usque ad Urbem conditam, dehinc usque ad Caesaris principatum natiuitatemque Christi ex quo sub potestate Urbis orbis mansit imperium, uel etiam usque ad dies nostros. . . .* Historians hold different opinions as to how far Orosius followed Augustine's concept of history and whether Augustine approved of Orosius' book: see T. Mommsen, "Orosius and Augustine," in *Medieval and Renaissance Studies,* ed. E. F. Rice (Ithaca, 1959), 325–48, and H. I. Marrou, "Saint Augustin, Orose et l'augustinisme historique," in *La storiografia altomedievale,* Settimane di Centro Italiano di studi sull'alto medioevo, 17 (Spoleto, 1970), 59–87.

8. On this pun and its idea in Roman literature, the writings of Orosius' contemporaries and his own, see the commentary in Orosius, *Le storie contro i pagani,* ed. A. Lippold (Milan, 1976), 1:367.

the world, as it were, through its various parts, burning with evils, set afire with a torch of greed, viewing them as from a watchtower...."⁹ By mentioning the watchtower *(specula)*, a place above the earth from which one can view its shape and events, Orosius evokes the tradition of contemplation, meditation on the imperfections and sins of the world—the tradition reflected in cosmic visions of saints and, probably, some medieval maps.¹⁰ By mentioning this, Orosius indicates how he is going to perceive and describe the world—in the same way Christ and his saints perceived it, that is, from a distance, as a generalized image representing the necessary background of human history.

The need to present this image, according to Orosius, is his reason for beginning with a geographical introduction: "first I shall describe the world itself that the human race inhabits, as it was divided by our ancestors [or predecessors] into three parts and then established by regions and provinces, in order that when the locale of wars and the ravages of diseases are described, all interested may more easily obtain knowledge not only of the events and times but also of places."¹¹ Thus Orosius concludes his statement of intent (which began with his own view of a connection between the earth and the people) by joining the old rhetorical tradition that requires historians describing events to report their time and place.

In the following section of his work, Orosius gives a long and fairly detailed geographical survey of the known world. He begins with its tripartite division into Asia, Europe, and Africa. Then he describes the provinces of the three continents, with their cities, rivers, and mountains. Orosius' geographical chapter contains only descriptions of the earth, omitting measurements.¹² Both in structure and contents, this survey conforms to the geographical knowledge of the time based on classical tradition. Orosius emphasizes this by beginning his survey with a reference to authorities, *maiores nostri.* More specifically, his chapter may go back to

---

9. Orosius *Hist.* I.1.14 (trans. p. 7): *. . . in quantum ad cognitionem uocare suffecero, conflictationes generis humani et ueluti per diuersas partes ardentem malis mundum face cupiditatis incensum e specula ostentaturus necessarium reor. . . .*

10. On the meaning of the term *specula* in the contemplative tradition and its connection to the cosmic visions and maps see Gautier Dalché, "De la glose," 755–56.

11. Orosius *Hist.* I.1.14 (trans. p. 7, modified by me): *. . . primum ipsum terrarum orbem quem inhabitat humanum genus, sicut est a maioribus trifariam distributus deinde regionibus prouinciisque determinatus, expediam; quo facilius, cum locales bellorum morborumque clades ostentabuntur, studiosi quique non solum rerum ac temporum sed etiam locorum scientiam consequantur.*

12. Ibid., I.2, pp. 9–40 in Zangemeister's *editio maior.*

the work of Marcus Vipsanius Agrippa, who, during the reign of Augustus, produced a description and a map of the world, now lost.[13]

The image of the world that emerges from Orosius' geographical chapter does not directly reflect any of the historical themes proclaimed in his statement of intent. This is an image of the earth seen from a watchtower—in its traditional classical outlines, it is generalized and timeless.[14] It provides a broad framework of reference for the following historical events, rather than time-specific topographical layout. Almost half of the cities that figure in the geographical chapter never appear again in the subsequent historical account.[15] Moreover, the majority of the geographical details introduced in the historical chapters are never mentioned in the geographical survey.[16] Neither of the emphases that Orosius makes within the classical picture of the world (he devotes more attention to certain regions and includes or omits certain details) seems to reflect a wish to fit his geography to his history.[17] Apparently, when he said that his geographical introduction would give the reader the knowledge of places as well as events and times,[18] he had in mind a general rather than a specific image. Concluding his geographical chapter, he emphasizes the transition from a general, ground-preparing survey of the whole world to a particular account of local and individual events.[19]

The general picture of the world given in Orosius' introduction is hard to localize in time. He provides no historical references for his placenames: the name of Carthage in the geographical chapter can signify both

---

13. Some historians even use Orosius' geographical chapter, along with two roughly contemporary works, the *Divisio orbis terrarum* and the *Dimensuratio provinciarum* (both edited in *Geographi latini minores*, 9–23) to reconstruct Agrippa's lost writings. See A. Klotz, "Beiträge zur Analyze des geographischen Kapitels in Geschichtswerk des Orosius (I.2)," in *Charisteria A. Rzach* (Reichenberg, 1930), 120–30; A. Lippold, however, suggests more caution in search for Orosius' sources, in Orosius, *Le storie*, 1:367–69.

14. Goffart (*Narrators*, 348) talks about the "timeless geography" of Orosius.

15. See Janvier, *La géographie*, 189. Out of the thirty-five cities mentioned in the geographical chapter, sixteen do not appear in the following account; see ibid., map 7.

16. For instance, the cities Sodom and Gomorra are only mentioned when Orosius recounts the biblical story (*Hist.* I.5.6); the Goths, who figure prominently in the history (they are mentioned twenty-two times; see the index in Zangmeister's *editio maior*, p. 344), do not appear in the geography.

17. On Orosius' emphases see Janvier, *La géographie,* chap. 7, "Les choix d'Orose."

18. Orosius *Hist.* I.1.17: *non solum rerum ac temporum sed etiam locorum scientiam consequantur.*

19. Ibid., I.2.106 (my emphasis): *Percensui breuiter ut potui prouincias et insulas orbis uniuersi. nunc locales gentium singularum miserias, sicut ab initio incessabiliter exstiterunt et qualiter quibusque exortae sunt, in quantum suffecero proferam.*

the great city once destroyed by Rome and the Roman Carthage contemporary to Orosius.[20] Several historians have pointed out that Orosius' geographical survey is full of anachronisms: sometimes the peoples he mentions had moved elsewhere or left the historical scene; most of the cities he names had long ago lost their importance.[21] This seems to result from his overall purpose rather than from poor knowledge or hasty use of unreliable sources.[22] Orosius does not want his picture of the world to be contemporary or even distinctly Christian. He evokes no significant biblical associations;[23] he mentions no biblical places: he even omits Jerusalem. Only once does he introduce a comparatively contemporary and Christian note, when he mentions Constantinople, previously called Byzantium.[24] All this seems to serve Orosius' purpose—to provide an image for the reader to contemplate, to create a physical background for future human events. This image had to be familiar enough, in its general classical outlines, that it would not distract the reader's attention from the main theme.

Orosius' geographical chapter is traditional in its contents. What seems new is its place within his work and within his concept of history. Although many historians before Orosius used geographical digressions of various length and some began their works with brief geographical expositions, no one had ever collected all geographical information and placed it in the beginning of his work, forming a long chapter.[25] Neither had any historian before Orosius made the image of the world serve as a contemplative background for the misfortunes of humanity.

Having initially indicated the connection between the earth and its inhabitants, Orosius, without referring to this matter in his geographical introduction, explains it later in his book. The earth and the whole physical world share the calamities of the human race as God's punishment for original sin: "This sentence of the Creator, God and Judge, destined for

---

20. Ibid., I.2.73; cf. V.1.5.

21. K. Müllenhoff, *Über die Weltkarte und Chorographie des Kaisers Augustus* (Kiel, 1856), 13–15. Yves Janvier gives a detailed study of Orosius' anachronisms, in *La géographie*, 226–37; on the cities see ibid., 189.

22. Contrary to previous opinions, Janvier (*La géographie,* esp. 141–43) has demonstrated that Orosius carefully adapted the information of his sources.

23. He actually mentions the Scripture once, when he says that it often calls the area from the Indus to the Tigris Media, at Orosius *Hist.* I.2.19: *sed generaliter Parthia dicitur, quamuis Scripturae Sanctae uniuersam saepe Mediam uocent.*

24. Ibid., I.2.56: . . . *Constantinopolim quae Byzantium prius dicta est.* . . .

25. On Orosius' innovations in concept and form see Corsini, *Introduzione,* "La nuova geografia"; Fabrini, *Paolo Orosio,* 322–27; Orosius, *Le storie,* 1:367; Janvier, *La géographie,* 9–10.

sinning men and for the earth because of men, and to last forever as long as men should inhabit the earth . . ."[26] Thus the earth, along with humans, enters the divine plan, reflected in the historical process. As predestined by God, the earth has a role of its own in human history.

This idea serves as a foundation for Orosius' concept of history and its progress: following God's ineffable plan, each part of the earth, in turn, becomes the place for one of the four great kingdoms—the Babylonian in the east, the Carthaginian in the south, the Macedonian in the north, and the Roman in the west[27]—and participates in their fate. Orosius emphasizes this by depicting the calamities of the human race on a universal scale, involving continents or even the whole world. He notes that two-thirds of the earth participated in the Greek and Persian wars: "all Asia and Europe, partly one against the other, partly among themselves, were intermingled with murders and shameful deeds."[28] Then, he writes that Rome conquered the earth and made it share her calamities: "Rome reckoned the extent of her kingdom by her misfortunes and, turning to her own destruction, laid claim to every individual nation in which she had conquered. To Asia, Europe, and Africa, I do not say to the three parts of the world but to every corner of these three parts, she exhibited her gladiators. . . ."[29] Finally, he reports that Rome, having "gained possession of Asia, Africa, and Europe"—that

---

26. Orosius *Hist.* I.3.2 (trans. p. 20): *sententiam creatoris Dei et iudicis peccanti homini ac terrae propter hominem destinatam semperque dum homines terram habitauerint duraturam.* . . . Cf. II.1.1: *unde etiam peccante homine mundus arguitur ac propter nostram intemperantiam conprimendam terra haec, in qua uiuimus, defectu ceterorum animalium et sterilitate suorum fructuum castigatur.* Cf. also I.5.11, where Orosius describes how God had punished even the earth of Sodom and Gomorra for the sins of their inhabitants: . . . *terra quoque ipsa, quae has habuerat ciuitates, primum exusta ignibus, post oppressa aquis, in aeternam damnationem communi periret adspectui.* On the doctrine of the original sin as an important component of Orosius' "new geography" see Corsini, *Introduzione,* 79–82.

27. Orosius *Hist.* II.1.5: . . . *ineffabili ordinatione per quattuor mundi cardines quattuor regnorum principatus distinctis gradibus eminentes, ut Babylonium regnum ab oriente, a meridie Carthaginiense, a septentrione Macedonicum, ab occidente Romanum.* . . . On the history of this idea and its development by Orosius see Corsini, *Introduzione,* 160–66; J. W. Swain, "The Theory of the Four Monarchies," *Journal of Classical Philology* 35 (1940): 1–21; Marrou, "Saint Augustin," 72–73; Fabrini, *Paolo Orosio,* 195ff.

28. Orosius *Hist.* II.18.3 (trans. pp. 73–74): *sic uniuersa Asia atque Europa, partim in se singulae, partim inter se inuicem funeribus et flagitiis permiscebantur.*

29. Ibid., VI.17.4 (trans. pp. 265–66): *Percensuit latitudinem regni sui Roma cladibus suis atque in suam conuersa caedem singulas quasque gentes ibidem, ubi domuit, uindicauit. Asiae Europae atque Africae, non dico tribus mundi partibus sed totis trium partium angulis edidit gladiatores suos.* . . .

is, the whole known world—became united under the power of one emperor, Augustus.³⁰

According to Orosius, the universal empire prepared the world for a universal religion. The united Rome brought peace to the earth, which helped to spread the Christian religion throughout the empire—that is, again, all over the world.³¹ Even the persecutions that in Orosius' history also reach universal proportions (the pagans search the whole earth to extinguish the Christian faith, and the blood of martyrs flows all over the empire) did not stop Christianity from conquering the world.³²

The earth also shares in human history by becoming an object of innumerable conquests. In Orosius' history, the power of an earthly kingdom is ultimately measured by the extent of its territory, and the rise of states is expressed in terms of gaining land. He writes that Carthage "surpassed all Africa and extended the boundaries of its empire, not only into Sicily, Sardinia, and other adjacent islands, but also into Spain."³³ He then notes that the Roman Empire, having already conquered three continents under Augustus, extended itself even more under Vespasian and Titus, including new provinces that obeyed Roman laws.³⁴ Finally, according to Orosius, the violation and devastation of the Roman lands by the barbarians, which he calls "the laceration of the Roman body," brought about the downfall of this great empire.³⁵

---

30. Ibid., VI.1.6 (trans. p. 229): . . . *postquam Asiae Africae atque Europae potitum est, ad unum imperatorem eundemque fortissimum et clementissimum cuncta sui ordinatione congessit.*

31. Ibid., VII.2.16: . . . *toto terrarum orbe una pax omnium non cessatione sed abolitione bellorum, clausae Iani geminae portae extirpatis bellorum radicibus non repressis census ille primus et maximus, cum in hoc unum Caesaris nomen uniuersa magnarum gentium creatura iurauit simulque per communionem census unius societatis effecta est;* VII.1.11: . . . *cuius [Christi] aduentui praedestinatam fuisse imperii Romani pacem.* . . . The idea that the establishment of the Roman Empire's dominion, granted by God, helped to overcome the discords and evils among the nations was also developed by Augustine in *De civitate Dei* V. On the dissimilarities between Augustine's and Orosius' concepts see Mommsen, "Orosius," 336–41.

32. Orosius *Hist.* VI.1.21: the persecutors say, . . . *Orbem totum scrutati sumus, siquo modo Christianum nomen et cultus uniuerso mundo posset abradi;* VII.22.3: . . . *fuso per omnem Romani regni latitudinem sanctorum sanguine.*

33. Ibid., VII.2.6 (trans. p. 286): . . . *Carthaginem uero uniuersae praecelluisse Africae et non solum in Siciliam Sardiniam ceterasque adiacentes insulas sed etiam in Hispaniam regni terminos tetendisse.* . . .

34. Ibid., VI.1.6: . . . *Asiae Africae atque Europae potitum est* . . . ; VII.1.7: . . . *Romanum imperium tam amplum ac tam sublime* . . . ; VII.9.10: . . . *in inmensum respublica Romana prouehitur: siquidem Achaia Lycia Rhodus Byzantium Samus Thracia Cilicia Commagene tunc primum redactae in prouincias Romanis iudicibus legibusque paruerunt.*

35. Ibid., VII.22.6: *soluuntur repente undique permissu Dei ad hoc circumpositae relictaeque gentes laxatisque habenis in omnes Romanorum fines inuehuntur;* VII.22.9 (trans. mine): *et ne quid forte Romani corporis ab hac dilaceratione cessaret, conspirant intrinsecus tyranni.* . . .

Among these global concerns, Orosius gives comparatively little attention to individual places. General trends and connections seem to interest him more than concrete locations. Describing the events of Eastern and Greek history, he uses a broad stroke more often than a minute touch, and he speaks of the conquests and movements of his characters in general terms of continents and provinces rather than in terms of specific cities or landscape features. Thus, he writes that the Amazons conquered "some cities in Asia"; that Cyrus, king of the Persians, overran "Asia, Scythia, and the entire Orient"; that during the Peloponnesian War, the Athenians "devastated the territory of the Spartans far and wide and added many cities of Asia to the Athenian empire."[36] Alexander the Great's conquests may be considered an exception—describing them, Orosius gives some geographical details, such as names of several conquered cities.[37] Turning to Roman history, Orosius' account becomes slightly more place specific; that is, he sometimes adds place-names. Thus, he gives the names of the towns in Latium captured by King Tarquinius Superbus; he specifies the farthest point of Roman conquests in Galatia in 189 B.C.; he mentions that during the barbarian invasions, the Germans, making their way over the Alps, through Raetia and all Italy, came as far as Ravenna.[38] Still, more often he never goes beyond the level of provinces; for instance, he only mentions the names of provinces when speaking of the invasions of other tribes.[39]

Orosius' rare and brief descriptions of locations largely demonstrate interests other than geographical. His description of Sodom and Gomorrah, located in the fertile valley of the Jordan River, prepares the ground for the account of God's judgment: the abundance of natural riches caused luxury and disgraceful passions, the latter leading to the destruction of the cities by heavenly fire.[40] The description of Babylon, emphasizing its mighty walls rather than the nature of its site, prepares Orosius' conclusion about the instability and changes in human history: "for whatever is

---

36. Ibid., I.15.5: ... *Asiae uero aliquantis ciuitatibus captis*...; II.6.1: ... *qui [Cyrus] tunc Asiam Scythiam totumque Orientem armis peruagabatur*...; I.21.15 (trans. p. 43): ... *Spartanorum fines late populati sunt et multas Asiae ciuitates Atheniensium imperio adiecerunt.* For other examples see his discussion of the Spartans' possessions (III.1.5–6) and of Philip of Macedon's conquests (III.13.4).

37. Ibid., III.16.11–12.

38. Ibid., II.4.12: ... *oppida ualida in Latio per eum [Tarquinium] capta Ardeam Oricolum Suessam Pometiamque*...; IV.20.25: *Fuluius consul de Graecia in Gallograeciam, quae nunc est Galatia, transuectus, ad Olympum montem peruenit*...; VII.22.7: *Germani Alpibus Raetia totaque Italia penetrata Rauennam usque perueniunt.*...

39. Ibid., VII.22.7: ... *Alamanni Gallias peruagantes etiam in Italiam transeunt; Graecia Macedonia Pontus Asia Gothorum inundatione deletur.*...

40. Ibid., I.5.6–11.

made by the hand and work of man collapses and is consumed by the passage of time, as the capture of Babylon confirms."[41] The same emphasis on walls and favorable strategic location prevails in the description of Carthage, eventually destroyed by Rome. Speaking of Carthage in his own time, Orosius creates an image of wretchedness, contrasting it to the former picture of greatness: "small in compass, destitute in walls," Carthage reminds people about the transience of mundane glory.[42] Pursuing his goal, Orosius seems to exaggerate the wretched condition of Carthage—from Augustine's *Confessions,* for instance, this city emerges as a vigorous center of intellectual and artistic life.[43]

Orosius makes both his descriptions of places and events serve his main goal—to demonstrate how divine providence shapes human history. This singularity of purpose may have contributed to the popularity of his book in the Middle Ages. Transmitted in multiple manuscripts and reflected in later writings, Orosius' view of historical and geographical questions became part of medieval thought.[44] Translated into Old English, Orosius' history began a new life, his geographical information updated and further transmitted in vernacular.[45]

Orosius' geographical chapter also acquired an independent existence and entered the geographical tradition on its own. It was sometimes transmitted separately from the rest of Orosius' book,[46] bearing a title of its

---

41. Ibid., II.6.7–11: the description of Babylon; II.6.13 (trans. p. 54): ... *quidquid est opere et manu factum, labi et consumi uetustate, Babylon capta confirmat.* ...

42. On the location and walls of Carthage see ibid., IV.22.4–7; on the transience of glory, V.1.5 (trans. p. 174): *cui [Carthagine] etiam nunc, situ paruae, moenibus destitutae, pars miseriarum est audire quid fuerit.*

43. Augustine *Confessiones* II.3, III.1–2, V.3. I thank Steven Epstein, who pointed out this connection to me.

44. On the manuscript tradition see J. M. Bately and D. A. Ross, "A Check List of Manuscripts of Orosius' *Historiarum adversum paganos libri septem," Scriptorium* 15 (1961): 329–34; they list 245 manuscripts, out of which 39 were produced before the eleventh century. On the influence of Orosius in medieval thought see Corsini, *Paolo Orosio,* 9–28; Corsini cites the previous literature.

45. See J. M. Bately, "The Relationship between Geographical Information in the Old English Orosius and Latin Texts other than Orosius," *Anglo-Saxon England* 1 (1972): 45–62. Bately also cites the previous literature.

46. Bately and Ross' checklist names the following pre-eleventh-century manuscripts as having only the geographical chapter: Munich, Bayerische Staatsbibliothek Clm 396, s. X, and Munich, Bayerische Staatsbibliothek Clm 29022, s. VIII, N. Italy. According to this checklist, other manuscripts contain excerpts from Orosius' history, but unfortunately it does not specify what chapters are excerpted. In addition, not listed in Bately and Ross is Albi 29, s. VIII, Spain or Septimania (*Mappa Mundi e codice Albigensi 29 accedunt Indeculum quod maria vel venti sunt et (Pauli Orosii) Discriptio terrarum ex eodem codice,* ed. F. Glorie in CCSL 175 pp. 467–87).

own: *Description of the Earth* or *The Account of the Whole Earth or Provinces.*[47] It became a school text[48] and an authority for later geographical writings.[49] The popularity of Orosius' geographical survey was not only due to his overall authority as a Christian historiographer—the fact that it represented a self-sufficient entity, in its contents only loosely connected to the historical text, has also made Orosius' geographical chapter a convenient tool.

Geographical concerns in Orosius' history represent an important part of his historical vision. Beginning with the survey of the world and throughout his book, the geographical themes—the earth as the object of contemplation of human misery, as a scene for human events, as a participant in these events, as a measure of mundane power—are interwoven with historical themes, the main one being the manifestation of divine providence in the vicissitudes of human history, eventually leading to the triumph of the true faith. Orosius' geographical view is more often directed toward the general than the specific, which is particularly obvious in the geographical introduction, generalized and timeless. Throughout his history, Orosius also often stays at the more general level.

Orosius' geographical introduction by itself established a model that proved to be very influential for both geographical and historical writings in the early Middle Ages. Geographers, borrowing from Orosius and combining his material with that going back to others, made Orosius' essentially classical image part of the medieval world picture and made his contribution part of medieval geographical knowledge. Historians placed geographical introductions in the beginnings of their books, borrowing the idea and sometimes the material from Orosius.

Jordanes

Jordanes, most probably working in Constantinople, wrote his *Gothic History* in the 550s as the third part of a historical compilation, also

---

47. *Discriptio terrarum,* Albi 29; *Liber canonum siue ratio totius orbis uel prouintiarum,* Munich, Bayerlische Staatsbibliothek Clm 396. In later manuscripts it is called *Cosmographia Orosii* (Florence Bibl. Laur. pl. 66, 19, s. XIV) or, descriptively, *Dicta Pauli Orosii historiographi de situ orbis, de Asia uidelicet maiore et minore, de Europa et de Africa* (Heidelberg, Pal. lat. 1568, s. XI–XII); see Zangemeister, "Die Chorographie des Orosius," 719.

48. In addition to the texts already mentioned, which were used at school, see, on glosses to Orosius' geography, Szerwiniack, "Un commentaire"; on a school text containing a dialogue on geography, based on Orosius and Isidore, in MS Paris, BN lat. 4892, see Gautier Dalché, *La "Descriptio mappe mundi,"* 97.

49. To name only some, Isidore's *Etymologies* in the seventh century, the *Cosmography* of Pseudo-Aethicus and the anonymous *De situ orbis terre uel regionum* in the eighth, and Dicuil's *De mensura* and the *De situ orbis* of Anonymus Leidensis in the ninth.

including the account of universal chronology and Roman history from Romulus to Jordanes' time.[50] Of the three parts, only *Gothic History* contains a geographical introduction; in addition, Jordanes makes numerous geographical digressions in the historical account.

Jordanes begins his geographical section immediately after a short preface, where he names his main source and his goal—to write an abbreviated version of "the twelve volumes of the Senator [Cassiodorus] on the origin and deeds of the Getae from olden times to the present day, descending through the generations of the kings."[51] The problem of Jordanes' relation to Cassiodorus' history has attracted much attention.[52] However, on the basis of the conjectures historians have made about Cassiodorus' lost work, it is impossible to decide how far Jordanes followed his main source in his treatment of geographical matters. Therefore, for my present purpose, I will concentrate on the existing text of Jordanes' history and interpret his geography as reflecting his goals and choices rather than those of Cassiodorus.

In the beginning of his book, Jordanes offers no explicit justification for his geographical introduction, which, nevertheless, has a clear goal that shapes its peculiar character. The goal and the culmination of Jordanes' geographical section is the description of Scandza, the island central to his story as the place of the Goths' origin. Jordanes himself points it out: "The same mighty sea [the northern sea] has also in its arctic region, that is, in the north, a great island named Scandza, from which my tale (by God's grace) shall take its beginning. For the race whose origin you ask to know burst forth like a swarm of bees from the midst of this island and came into the land of Europe."[53]

Jordanes builds his geographical introduction as a frame for his account of Scandza. To achieve this, he changes the classical format of a geographical description, which usually listed first the continental

---

50. Jordanes, *Romana et Getica,* ed. T. Mommsen, MGH AA V (Berlin 1882), 1–138. Unless otherwise noted, translations of Jordanes' *Getica* are from *The Gothic History of Jordanes,* trans. C.C. Mierow. (2d ed., Princeton, 1915; reprint, Cambridge and New York, 1966). For a summary of research on Jordanes' biography, his history, and its sources see Goffart, *Narrators,* chap. 2.

51. Jordanes *Getica* 1 (trans. p. 51): . . . *suades, ut nostris verbis duodecem Senatoris volumina de origine actusque Getarum ab olim et usque nunc per generationes regesque descendentem in uno et hoc parvo libello choartem.* . . .

52. See the summary of the discussion in Goffart, *Narrators,* 23–42.

53. Jordanes *Getica* I.9 (trans. p. 53): *habet quoque is ipse inmensus pelagus in parte artoa, id est septentrionali, amplam insulam nomine Scandzam, unde nobis sermo, si dominus iubaverit, est adsumpturus, quia gens, cuius originem flagitas, ab huius insulae gremio velut examen apium erumpens in terram Europae advinit.* . . .

provinces and then the islands.⁵⁴ He devotes only two sentences to the tripartite division of the earth and the provinces, briefly lists the islands of the southern and western parts of the ocean, and discourses at length about two northern islands, Britain and Scandza, giving each as much space as the rest of the earth.⁵⁵

Thus freely shaping the classical plan according to his own needs, Jordanes at the same time demonstrates a great concern to fit into it. He begins his geographical introduction in the same words as Orosius had begun his: "Our ancestors, as Orosius relates, were of the opinion that the circle of the whole world was surrounded by the girdle of Ocean. . . . Its three parts they called Asia, Europe, and Africa."⁵⁶ This reference to the authority of Orosius is all the information that Jordanes gives the reader on the tripartite division of the earth. In the following sentence, Jordanes disposes of the necessity to describe the provinces or the measurements of the world by citing additional authorities who have already done this: "there are almost innumerable writers, who not only explain the situations of cities and places, but also measure out the number of miles and paces to give more clearness."⁵⁷ Thus referring his reader to information that must have been very familiar and easy to obtain, Jordanes also establishes the traditional framework for his following account.

He does the same in his list of the islands: it is supported by the authority of Orosius, and their order largely conforms to that of the previous tradition.⁵⁸ The same is true for Jordanes' account of Britain: he has Pompo-

---

54. Roman geographical writers, such as Pliny and Pomponius Mela, followed the scheme of *periplus* (the description of a sailing route along the coast) and treated first the coasts and interior regions, then the islands situated near the coasts. Orosius, too, used this scheme.

55. On the three parts of the world and for the list of islands see Jordanes *Getica* I.4–9 (pp. 54–55 of Mommsen's edition); on Britannia, II.10–15 (pp. 56–57); on Scandza, III.16–23 (pp. 58–59.)

56. Ibid., I.4 (trans. p. 52): *Maiores nostri, ut refert Orosius, totius terrae circulum Oceani limbo circumseptum triquadrum statuerunt eiusque tres partes Asiam, Eoropam et Africam vocaverunt* (= Orosius *Hist.* I.2.1). For discussion of the term *triquadrus* see M. Smyth, *Understanding the Universe in Seventh-Century Ireland* (Woodbridge, 1996), 280.

57. Jordanes *Getica* I.4 (trans. p. 52): *de quo trepertito orbis terrarum spatium innumerabiles pene scriptores existunt, qui non solum urbium locorumve positiones explanant, verum etiam et quod est liquidius, passuum miliarumque dimetiunt quantitatem.* . . .

58. As Mommsen's parallel shows, Jordanes' list of southern islands follows that of Julius Honorius, the fourth- or fifth-century cosmographer; cf. Jordanes *Getica* I.6 and Julius Honorius *Cosmographia* 3. Svennung suggests that Jordanes may have used a map similar to the one that accompanied the text of Julius Honorius; see Svennung, *Jordanes und Scandia*, 9, and Dagron, "Une lecture," 297 n. 2.

nius Mela, Tacitus, and others to rely on.[59] However, Scandza, the real focus of Jordanes' geographical introduction, was not that well attested in the sources available to him.

In the beginning of his account about Scandza, he quotes Ptolemy, "an excellent describer of the world," who mentions "a great island situated in the surge of the northern Ocean, Scandza by name, in the shape of a juniper leaf."[60] Ptolemy does mention Scandia, the largest of the four islands located near the mouth of the Vistula.[61] However, Mela, whom Jordanes also cites, only talks about *Codanus sinus* (located somewhere in the southern part of the Baltic Sea), never mentioning Scandza.[62] Jordanes probably only names Mela as the closest reference on Scandza that he could find. As a whole, the support classical authority could offer on the question of Scandza appears rather slim.

It seems impossible to decide where Jordanes got the details for his description of Scandza—that it has in its eastern part a vast lake, whence the river Vagus flows into the ocean; that it has many small islands scattered around it; that many nations dwell there, with their own customs; and so forth.[63] He may have relied on a written Gothic tradition, such as the history of Ablavius, who is identified by Jordanes as "a famous describer of the Gothic race."[64] That Jordanes describes him in the same words he had earlier applied to Ptolemy—"an excellent describer of the world"—may imply that in referring to their "descriptions" he meant both historical and geographical ones. However, Jordanes never mentions

---

59. See the parallels in Mommsen's edition, pp. 56–57.

60. Jordanes *Getica* I.3.16 (trans. p. 55): *de hac [Scandza] etenim in secundo sui operis libro Claudius Ptolomeus, orbis terrae discriptor egregius, meminit dicens: est in Oceani arctoi salo posita insula magna, nomine Scandza, in modum folii cetri.* . . . Svennung (*Jordanes und Scandia*, 9) insists that the form of this island resembled a citrus, rather than juniper, leaf, according to the reading *citri* (citrus) rather than *cetri* (or *cedri*, "juniper") found in some manuscripts.

61. Ptolemy *Geographia* II.11.33–35.

62. Pomponius Mela, *De situ orbis*, ed. C. Frick (Leipzig, 1880), III.3.31. Elena Skrzhinskaya thinks this is the Gdansk Bay; see Iordan, *O proishozhdenii i deyaniyah Getov: Getica*, (On the Origin and Deeds of the Getae: Getica) ed. and trans. E. Ch. Skrzhinskaya (Moscow, 1960), 190 n. 44.

63. Jordanes *Getica* III.17–24. Weibull ("Scandza und ihre Völker," esp. 218–24) thinks that this information comes from the classical sources, garbled by Jordanes. Svennung (*Jordanes und Scandia*, esp. 14–28) argues that Jordanes' information comes from eyewitnesses—travelers and merchants—and describes real places that can be identified. Herwig Wolfram also suggests that Jordanes records geographical knowledge, current and valid in the sixth century; see H. Wolfram, *History of the Goths*, trans. T. J. Dunlap (Berkeley and London, 1988), 38–39.

64. Jordanes *Getica* IV.28: . . . *Ablavius descriptor Gothorum gentis egregius*.

Ablavius as an authority on Scandza, only naming him later when talking about the migration of the Goths to Scythia.

It is less likely that Jordanes used the writings of the Gothic geographers who supplied some information for the eighth-century Anonymus Ravennatis.[65] This cosmographer, our only source that mentions the lost Gothic geographies, turns directly to Jordanes' account of Scandza without naming any other writers.[66]

Could the description of Scandza go back to some kind of oral tradition—either the eyewitness reports of travelers and merchants[67] or the Gothic "ancient songs" and legends that, according to Jordanes, tell the story of the Gothic migrations "in almost historical fashion"?[68] This would be a tempting solution; however, it runs into the same difficulty as the attribution of the information to Ablavius—Jordanes only mentions both in connection with the Gothic migration rather than as sources for the story of Scandza.

Whether the major part of Jordanes' description of Scandza came from a written or an oral tradition, Jordanes may have felt that, on its own, it lacked the support of authorities; thus, he tried to supply even remotely appropriate references. In a similar way, he hastened to support the "almost historical" evidence of the "ancient songs" by the authority of both Ablavius, who Jordanus says "confirms this in his most trustworthy account," and more ancient writers (called *maiores,* just like those in the first sentence of the geographical introduction), who "also agree with the tale." Here Jordanes names only Josephus Flavius, "a most reliable relator of annals," who actually omits the story of the Gothic origins and only mentions the Scythians.[69] On the same shaky grounds, Jordanes earlier named Pomponius Mela as an authority on Scandza.

---

65. The Gothic geographers' works mentioned by Anonymus Ravennatis are lost. On their possible dates and contents see Staab, "Ostrogothic Geographers." For a different point of view see Dillemann, *La Cosmographie,* 57–58.

66. Anonymus Ravennatis *Cosmographia* I.12, p. 11: *quam insulam plerique phylosophi . . . historiographi conlaudant; quam et Iordanus, sapientissimus cosmographus, Scanzan appellat. ex qua insula . . . pariterque gentes occidentales egresse sunt. . . .*

67. As suggested by Svennung, *Jordanes und Scandia,* 14–28.

68. Jordanes *Getica* IV.28 (trans. p. 58) : *quemadmodum et in priscis eorum carminibus pene storicu ritu in commune recolitur. . . .*

69. Ibid., IV.28–29 (trans. p. 58): *quemadmodum et in priscis eorum carminibus pene storicu ritu in commune recolitur: quod et Ablavius descriptor Gothorum gentis egregius verissima adtestatur historia. in quam sententiam et nonnulli consensere maiorum: Ioseppus quoque annalium relator verissimus dum ubique veritatis conservet regulam et origines causarum a principio revolvat. haec vero quae diximus de gente Gothorum principia cur omiserit, ignoramus: sed tantu Magog eorum stirpe comemorans, Scythias eos et natione et vocabulo asserit appellatos.*

This parallelism in the methods of proof might suggest certain parallelism in the nature of sources in both cases—like the whole migration story, the account of Scandza might ultimately go back to the Gothic tradition, probably transmitted in oral form. Like the migration story (or as a subsection of it), the geographical information, in the eyes of Jordanes, may have needed particular proof. Summing up his discussion of the origin of the Goths, he admits his trust in written rather than oral tradition: "For myself, I prefer to believe what I have read, rather than put trust in silly tales."[70] Although this cannot be taken as a pure declaration of method (Jordanes means here a specific "silly tale" about the British origins of the Goths[71] rather than the oral tradition in general), the way he proceeds to refute it suggests that he considers written evidence more weighty than the oral. According to Jordanes, we read about Goths dwelling in Scythia, but there is nothing in their written records about Britain—which confirms the silliness of the tale.[72]

Jordanes' concern with authorities may have been among the considerations shaping the structure and contents of his geographical introduction, which, like Orosius' survey, includes information that will never appear in the historical account. Though the names of the three continents mentioned in passing will figure from time to time,[73] the list of islands and Britain will not. Jordanes may have focused on these islands rather than the dry land to create a coherent classical frame that would extend its authoritativeness to Scandza, introduced as one island among many and seemingly supported by tradition.[74] This consideration may in part explain why Jordanes introduced his long description of Britain, well known from classical sources but rather irrelevant for the subsequent story of the Goths except as a possible place of their origin.

---

70. Ibid., V.38 (trans. p. 60, modified by me): . . . *nos enim potius lectioni credimus quam fabulis anilibus consentimus.*

71. On this see Goffart's ingenious explanation of Jordanes' treatment of "Britain and the laborious invention of a Scandzan alternative" as a response to the political situation of A.D. 538 when the Goths were offered Britain as the place for their settlement (*Narrators,* 91–96).

72. Jordanes *Getica* V.38: *quorum mansione prima in Scythiae solo iuxta paludem Meotidem . . . legimus habitasse: nec eorum fabulas alicubi reperrimus scriptas, qui eos dicunt in Brittania vel in unaqualibet insularum in servitute redactos . . . aut certe si quis eos aliter dixerit in nostro urbe, quam quod nos diximus, fuisse exortos, nobis aliquid obstrepebit: nos enim potius lectioni credimus quam fabulis anilibus consentimus.*

73. Asia is mentioned six times, Africa eight times, Europe never outside geographical descriptions.

74. Dagron ("Une lecture," 298–99) suggests that Jordanes needed the account of the islands (and the whole classical geographical framework) to introduce the new people—the Goths—into the Roman world.

To support his description of Scandza, Jordanes makes it sound parallel to that of Britain.[75] Reporting that both islands are large and are situated in the northern sea, he notes that numerous Greek and Latin authors wrote on the position and shape of Britain,[76] while the same information on Scandza has not been traced back to any known source. He describes both islands as rather unattractive, writing that Britain is surrounded by the sluggish sea and "exhales such mists from its soil, soaked by the frequent inroads of Ocean, that the sun is covered throughout the whole of their disagreeable sort of day. . . ." He adds that its products, although abundant, can feed beasts rather than men, and he writes that its people are wild and live like beasts.[77] He reports that Scandza is also surrounded by an immense ocean, frozen in winter, and that the sea is unnavigable in the north. He concludes that "the land is not only inhospitable to men but cruel even to wild beasts," and that its people, like the population of Britain, resemble beasts by their cruelty in fighting.[78]

Thus Jordanes establishes Scandza as one place among others whose description follows the classical tradition. Does it also come to share in that tradition's timelessness? All Jordanes' information about the continents and islands is so general that it is hard to locate in time. The same is true for his descriptions of Britain and Scandza. Britain, described in terms of Strabo, Tacitus, and Dio Cassius, belongs to the pre-Christian period.[79] On Scandza and its peoples, as Jordanes himself points out, he gives information more recent than Ptolemy's.[80] The findings of those scholars who have worked on identifying the nations mentioned by Jor-

---

75. This parallelism has been often noticed by historians. Dagron ("Une lecture," 297–98) considered it "une fausse symmetrie" justifying the transition from the real island of Britain to the mythical island of Scandza. Goffart (*Narrators*, 91) pointed out Jordanes' comparison of Britain and Scandza as an elaboration of the Scandzan alternative.

76. Jordanes *Getica* II.10. As Mommsen has shown, Jordanes' information goes back to Tacitus, Mela, Strabo, and Dio Cassius (see Mommsen's edition, p. 56, nn. 2–4, p. 57, nn. 1–2.)

77. Jordanes *Getica* II.10–15 (trans. pp. 53–55), II.12: . . . *tantas illam exalare nebulas, madefacta humo Oceani crebris excursibus, ut subtectus sol per illum pene totum fediorem, qui serenus est, diem negetur aspectui.*

78. Ibid., III.16–24 (trans. pp. 55–57), III.18: *ita non solum inhospitalis hominis, verum etiam beluis terra crudelis est.*

79. Christianity does not enter Jordanes' picture of Britain. Goffart (*Narrators*, 90 n. 335) points out that the scythed chariots described at *Getica* II.15 indicate pre-Roman peoples: they fight *bigis curribusque falcatis, quos more vulgare essedas vocant.*

80. Pointing out at *Getica* III.19 that Ptolemy only mentions seven nations, he proceeds to record much more: *Scandza vero insula . . . licet multae et diversae maneant nationes, septem tamen eorum nomina meminit Ptolemaeus.*

danes show that at least by the end of his Scandza account, he records the situation close to his own time.[81] In the rest of the book, Jordanes' geography also seems to be that of his time: he gives contemporary information on the settlement of the tribes in Scythia; he records geographical names current in the sixth century.

Throughout Jordanes' book, geography serves his main goal, the story of the Goths. His geographical descriptions highlight the places important in Gothic history; that he frames the longest of them as digressions serves to attract the reader's attention. This use of geographical digressions fits the rhetorical tradition—to describe locations and prepare the scene before talking about events and people. Jordanes specifically introduces his section on Britain, important as a parallel to Scandza: "But now I shall speak briefly as I can about the island of Britain." And he ends it by a transition to Scandza, to emphasize the parallelism: "Let it suffice to have said these few things on the shape of the island of Britain. Let us now return to the site of the island of Scandza, which we left above."[82] In the same way, Jordanes emphasizes the importance of Scythia, the Goths' destination after they left Scandza. Interrupting the account of the Goths' migration, he introduces his long geographical description: "Before we enter on our history, we must describe how the boundaries of this land lie."[83]

In general, places and their locations seem to be important to Jordanes inasmuch as they are connected to the Goths. At least, only such places get brief descriptions, from simple indication of boundaries to a minireport on the location, products, and peoples.[84] Places famous because of the Gothic past can even serve as landmarks—in one case, Jordanes localizes the capi-

---

81. Ibid., III.24, on Roduulf, the king of the Ranii who fled to Theodoric the Great; see Svennung, *Jordanes und Scandia,* 182–83, and Goffart, *Narrators,* 90.

82. Jordanes *Getica* II.10 (trans. p. 53): *Nunc autem de Britannia insula... ut potuero, paucis absolvam;* II.15–III.16 (trans. p. 55): *haec pauca de Britanniae insulae forma dixisse sufficiat. Ad Scandziae insulae situm, quod superius reliquimus, redeamus.* The translations are modified by me.

83. Ibid., IV.29 (trans. p. 58, modified by me): *cuius soli terminos, antequam aliud ad medium deducamus, necesse est, ut iacent, edicere.* The description of Scythia takes two and a half pages in Mommsen's edition, approximately as much as the description of Scandza. Other long geographical digressions in Jordanes are the discussions of the Caucasus in the story of the Gothic Scythian-born Amazons (VII.53–55); the Danube, the southern boundary of the Goths in Dacia (XII.75); and Ravenna, the future capital of the Goths (XXIX.148–51).

84. For instance, ibid., IX.59, the description of Moesia, the kingdom occupied by the Goths in the times of Hercules, which goes back to Orosius (cf. Orosius *Hist.* I.2.55); XII.74, on Dacia, which the Gepidae, a branch of the Goths, possessed at the time of Jordanes; XXX.156, on the land of the Bruttii at the south of Italy, the place of Alaric's death.

tal of Attila the Hun as standing near the place "where long ago Vidigoia, bravest of the Goths, perished by the guile of the Sarmatians"—an indication that he seems to have added to the information borrowed from his source.[85] Thus, Jordanes consistently makes his geography serve his history.

In his geographical description, Jordanes established a tradition that formed part of early medieval geographical knowledge. In the eighth century, Anonymus Ravennatis, in his *Cosmography,* called Jordanes "the wisest cosmographer," borrowed from his account of Scandza, and referred to his opinion in other instances.[86] In the ninth century, Freculf of Lisieux, in his world chronicle, copied Jordanes' description of Scandza almost verbatim.[87] Jordanes' becoming an authority in geographical matters may have resulted from his own regard for classical authority. Having put much effort into making his geographical descriptions appear as a part of established classical tradition, he succeeded in establishing his trustworthiness in the eyes of early medieval scholars.

In his introductory survey, centered on the place of the Gothic origin, and in his discussion of places important for the story of the Goths, Jordanes' goals shape his vision of space as his history develops. He demonstrates a great concern to place this vision in the framework of classical tradition, supporting what he may have learned from oral sources by references to well-known authors. The classical framework created by Jordanes is as generalized and timeless as the geography in Orosius' introduction. However, the rest of his information seems contemporary and specific, serving to emphasize and clarify the events in Gothic history.

Bede

Bede defines the main goal of his book in its first sentence: it is to provide "the ecclesiastical history of the English nation."[88] As historians often

---

85. Ibid., XXXIV.178 (trans. p. 101): *venimus in loco illo, ubi dudum Vidigoia Gothorum fortissimus Sarmatum dolo occubuit.* Jordanes' source is Priscus, a fifth-century historian; see Mommsen's edition of Jordanes, pp. xxxiv–xxxv and p. 104, n. 3. The reference to Vidigoia is absent from the surviving fragments of Priscus; see *The Fragmentary Classicising Historians of the Later Roman Empire,* ed. and trans. R. C. Blockley, 2 vols. (Liverpool, 1981–83), 2:261.

86. Anonymus Ravennatis *Cosmographia* I.12, p. 11: . . . *quam [insulam] et Iordanus, sapientissimus cosmographus, Scanzan appellat.* For the discussion of Jordanes as a source for Anonymus Ravennatis see J. Schnetz, "Jordanis beim Geographen von Ravenna," *Philologus* 81 (1925–26): 86–100.

87. Freculph of Lisieux, *Chronicon,* PL 106, 961B–C, I.2.16.

88. Bede *EH, Praefatio,* p. 3 (trans. modified by me): *Historiam gentis Anglorum ecclesiasticam, quam nuper edideram.* . . .

pointed out, Bede understood this as the story of Christianity in England developing within the divine plan and reflecting the work of the divine providence.[89] What was the role of places in this "sacred history"?

In the preface to the *Ecclesiastical History,* Bede immediately indicates his interest in localizing the human events he chose to write about. He notes that he wished "to put on record concerning each of the kingdoms and the more important places, those events which I believed to be worthy of remembrance and likely to be welcome to the inhabitants. . . ."[90] He also takes special care to localize his sources; earlier in the same preface, he explains in detail where his information about particular places (Kent, East Anglia, Northumbria, Mercia, etc.) came from.[91] By doing this, Bede lays out a kind of preliminary geographical survey and immerses his reader in the eighth-century world of English political and ecclesiastical divisions. These are the first indirect indications of the importance that Bede attached to places in his history. The next one is the fact that he begins it with a geographical introduction, immediately following the preface.

Bede offers no explicit justification for his description of Britain. He may have felt, like other historians before him, that the tradition of a geographical introduction to a historical work had been sufficiently established. Orosius' history, which Bede knew, provided an ample justification and established an example to follow. Gildas, whose history Bede used, began with the description of Britain without justifying it, but simply including it in the list of his themes.[92]

Bede links his geographical account to this tradition by locating Britain in the classical picture of the world. To do that, Bede gives a skillfully arranged string of quotations, each so brief that it is hard to attribute them

---

89. On Bede's goal and the traditions in which it stands see Hunter Blair, *Bede's Ecclesiastical History;* Ray, "Bede, the Exegete"; Wallace-Hadrill, *Bede's Ecclesiastical History,* 2–3.

90. Bede *EH, Praefatio,* p. 7 (trans. modified by me): *qui de singulis prouinciis siue locis sublimioribus, quae memoratu digna atque incolis grata credideram, diligenter adnotare curaui....*

91. Bede *EH, Praefatio.* Much has been written on Bede's sources. For a general outline and bibliography see Colgrave, "Historical Introduction," xxx–xxxiv, in his and Mynors' edition of *EH;* Wallace-Hadrill, *Bede's Ecclesiastical History,* 3–5.

92. Gildas *De excidio* 2, pp. 27–28: *Sed ante promissum deo volente pauca de situ, de contumacia, de subiectione . . . dicere conamur.* The description appears at ibid., 3, pp. 28–29. Colgrave's suggestion (in his and Mynors' edition of *EH,* p. 14, n. 1) that Bede may have borrowed the idea of a geographical introduction from Gregory of Tours as well as Orosius seems far-fetched. The *History of the Franks* begins with an account of creation rather than a geographical survey.

definitively to a single source.[93] The first sentence, in which Bede defines the location of Britain—an island of the ocean lying "opposite Germany, Gaul, and Spain, which form the greater part of Europe, though at a considerable distance from them"—goes back to Orosius and Pliny.[94] The second sentence, about the size of Britain, Bede borrows from Orosius, Gildas, and Solinus.[95] The third, further specifying the location of Britain, quotes Orosius and Pliny. Orosius' passage on the Orkney Islands lying to the west of Britain concludes the borrowed passage.[96] This display of classical authority, probably well known to medieval readers, both creates the context for and lends support to the rest of the geographical chapter, mainly Bede's own.[97]

Bede makes his brief borrowed passage very functional. He immediately begins with England, the focus of his attention. The few places he mentions—Germany, Gaul, Spain, the ports on both sides of the Channel—are relevant for the subsequent historical account. They will either figure there directly or be connected to the events in some way.[98] Bede also introduces a distinctly English note into the classical tradition, by recording the local equivalent of the Latin name *Rutubi Portus: Reptacaestir* (Richborough).[99] As a result, the borrowed passage, even though it goes back to authorities, sounds slightly less classical and timeless than Orosius' or Jordanes' geographical introductions. However, Bede may not have aimed at this particular impression; he only carefully arranges his

---

93. Bede *EH* I.1; the borrowed passage occupies approximately one-fourth of the whole geographical introduction.

94. Ibid., I.1: <u>Britannia</u> Oceani <u>insula, cui quondam Albion nomen fuit, inter septentrionem et occidentem</u> locata est, <u>Germaniae, Galliae, Hispaniae, maximis Europae partibus, multo interuallo aduersa.</u> The emphasis here follow Plummer's edition, indicating Bede's borrowings. The beginning of the first phrase seems to go back to Orosius, the rest of it to Pliny *Nat. hist.* IV.30.

95. Orosius *Hist.* I.2.77; Gildas *De excidio* 3, p. 28; Solinus, *Collectanea rerum memorabilium*, ed. T. Mommsen (Berlin, 1895), XXII.10.

96. Bede *EH* I.1: *A tergo autem, unde Oceano infinito patet, Orcadas insulas habet* (= Orosius *Hist.* I.2.78).

97. At least, most of it has not been traced back to authorities; see Colgrave's and Mynor's edition of *EH*, p. 14, n. 1.

98. Germany as the place of origin of various peoples coming to England (Bede *EH* I.15, V.9); Spain as the country ruled by Emperor Constantius, who died in Britain (I.8); Gaul, most frequently, as the region connected with Britain by missionary activity. For the numerous references to Gaul see the index in Bede, *Venerabilis Baedae opera historica*, 2:465–66. The crossing of the Channel is mentioned in connection to the travels from and to Gaul; see, for instance, Caesar's crossing in Bede *EH* I.2.

99. Bede *EH* I.1: . . . *ciuitas quae dicitur Rutubi Portus, a gente Anglorum nunc corrupte Reptacaestir uocata.* . . .

authorities and truthfully records an English place-name lacking in the classical Graeco-Roman geographical tradition.[100]

The lack of detailed descriptions of Britain in the classical geographical tradition may have been the reason why Bede supplemented and corrected his authorities later in his book. Quoting Orosius on Caesar's conquest of Britain, he omits Orosius' erroneous statement that the Thames is only fordable at one point.[101] He also interrupts the quotation to observe that the traces of the stakes that had blocked the bank of the river and the ford from the Romans can still be visible in his own time.[102] Thus, he at the same time updates Orosius and gives additional evidence. In the other case, talking about Vespasian's conquests in Britain and following Eutropius' history of the Romans (written ca. 369),[103] Bede inserts some information about the Isle of Wight—its size and distance from the southern coast of Britain—between the quotations from Eutropius.[104] Here, as in his geographical introduction, Bede keeps in mind his main focus, Britain, and accordingly complements his authorities.

The classical image of Britain does not occupy Bede's entire introduction. Having rendered his tribute to the classical tradition, Bede turns to another one—that of the praise of land and paradisiacal settings. Immediately after the borrowed passage, he begins his praise of Britain and Ireland, describing both as abundant in their natural riches and exceptionally healthy and beneficial for human habitation. England, in Bede's description very close to Paradise, bears no resemblance to the utterly unpleasant island of the same name depicted by Jordanes. Bede pursues quite a different goal; evoking the biblical images of Paradise and the Promised Land, with all the exegetical levels of meaning behind them, he creates the setting for his story, which also contains several levels of meaning. Bede writes about the work of divine providence as manifested in the journey of the English nation toward Christianity—from their prefallen state, through

---

100. On the importance of place-names in Bede's method see J. McClure, "Bede's Old Testament Kings," in *Ideal and Reality in Frankish and Anglo-Saxon Society,* ed. P. Wormald (Oxford, 1986), 76–129.

101. Cf. Bede *EH* I.2, and Orosius *Hist.* VI.9.6: <u>inde ad flumen Tamesim profectus, quem uno tantum loco uadis transmeabilem ferunt. in huius ulteriore ripa Cassouellauno duce immensa hostium multitudo consederat ripamque fluminis ac paene totum sub aqua uadum acutissimis sudibus praestruxerat.</u> In Orosius' text, Bede's borrowings are emphasized following the convenient practice of Plummer's edition. On Bede's omission see Plummer's note *in Venerabilis Baedae opera historica,* 2:13.

102. Bede *EH* I.2.

103. Eutropius, *Breviarium ab urbe condita,* ed. H. Droysen, MGH AA II (Berlin, 1961).

104. Bede *EH* I.3. Cf. Eutropius *Breviarium* VII.19 and VII.14.

their fall, to their salvation.[105] In this picture, the paradisiacal setting, evoking the life before original sin, symbolizes the same state of Britain. By implication, people who live in this blessed place possess certain special characteristics that make them fit to receive conversion. Bede sees as another sign of divine providence the five languages spoken in Britain—English, British, Irish, Pictish, and Latin—which correspond in their number to the five books of divine law, the Pentateuch, "all devoted," Bede writes, "to seeking out and setting forth one and the same kind of wisdom, namely the knowledge of sublime truth and of true sublimity."[106] He argues that just as the five books of the Scripture are directed toward the supreme wisdom, so the people living in Britain, in all their languages, are predestined to seek the wisdom of Christianity.[107]

Certain special characteristics later in Bede's history directly lead to the conversion of the English. Before telling his famous story of how Gregory the Great decided to send a mission to England, Bede explains that he has included this tale, which goes back to popular tradition, because it gives the reason why Gregory "showed such earnest solicitude for the salvation of our race."[108] According to the story, when Gregory saw English slaves, fair and handsome, put out for sale, he asked where they came from and whether they were Christians. On hearing that they were not, he expressed his regret that men so bright of face should be in the power of the Prince of Darkness. Then, when he heard that their nation was called the Angli, their kingdom Deiri, and their king Aelle, he replied by three puns, playing on the similarly sounding words: "they have the face of angels *[angelicam]*, and such men should be fellow-heirs of the angels *[angelorum]* in heaven"; "*De ira!* Good! Snatched from the wrath of Christ and called to his mercy"; and "Alleluia! The praise of God the Creator must be sung in those parts."[109] In this story, as told by Bede, certain special features that

---

105. On Bede's introduction see Kendall, "Imitation," 178–82, and Wallace-Hadrill, *Bede's Ecclesiastical History*, 6, 11. On Bede's concept of history and his use of Gildas' similar ideas see Hanning, *Vision of History*, 75–90.

106. Bede *EH* I.1: *Haec in praesenti iuxta numerum librorum quibus lex diuina scripta est, quinque gentium linguis unam eandemque summae ueritatis et uerae sublimitatis scientiam scrutatur et confitetur, Anglorum uidelicet Brettonum Scottorum Pictorum et Latinorum. . . .*

107. This general vision does not prevent Bede from seeing and sharply criticizing the faults of Britain's inhabitants: on this see Goffart, *Narrators*, 302.

108. Bede *EH* II.1: *Nec silentio praetereunda opinio, quae de beato Gregorio traditione maiorum ad nos usque perlata est, qua uidelicet ex causa admonitus tam sedulam erga salutem nostrae gentis curam gesserit.*

109. Ibid., II.1, pp. 133, 135: *. . . nam et angelicam habent faciem, et tales angelorum in caelis decet esse coheredes. . . . Deiri, de ira eruti et ad misericordiam Christi uocati. . . . Alleluia, laudem Dei Creatoris illis in partibus oportet cantari.*

make Gregory notice the English slaves—physical beauty and particularly names—eventually serve their salvation.[110] Thus Bede's description of the people continues the theme begun in his description of the land, supporting his main goal.

This goal, to tell the story of conversion, also influences the way Bede describes places throughout his history. Not only does he mention places relevant to the progress of Christianity, but he indicates their location, naming the province, the city, or distance from it, and sometimes giving topographical details, such as rivers or woods close to it. For Bede, it is important to record not only how and when but where people get baptized. King Edwin of Northumbria received baptism at York, in the Church of St. Peter the Apostle;[111] afterward, Paulinus baptized many Northumbrians in the kingdom of Bernicia in the river Glen and in the kingdom of Deira "in the river Swale which flows beside the town of Catterick."[112] A certain old man recalled that he had been baptized by Paulinus "in the river Trent, near a city which the English call *Tiowulfingacaestir* [Littleborough]."[113] The Isle of Wight, in Bede's story the last part of Britain to be converted, even gets a second description following the account of its conversion—a paragraph describing its location and the ocean tides that flow around it.[114]

The location of the newly founded monasteries also attracted Bede's attention. The monastery founded by Augustine was located "not far from the city [Canterbury], to the east."[115] Fursa, a holy man from Ireland, built a monastery "close to the woods and the sea, in a Roman camp which is called in English *Cnobheresburg,* that is, the city of Cnobhere [Burgh Cas-

---

110. For Bede natural beauty can be a special sign of grace: so, the beauty of the place where St. Alban died makes it fit to be hallowed by the martyr's blood (ibid., I.7): . . . *Natura complanat, dignum uidelicet eum pro insita sibi specie uenustatis iam olim reddens, qui beati martyris cruore dicaretur.* On the deep meaning of the names of the Angli cf. the account in *Life of Gregory* written at Whitby (*The Earliest Life of Gregory the Great, by an Anonymous Monk of Whitby,* ed. and trans. B. Colgrave [Lawrence, Kans., 1968], 12–13, pp. 94–96).

111. Bede *EH* II.14: *Baptizatus est autem Eburaci die sancto paschae pridie iduum Aprilium, in ecclesia sancti Petri apostoli.* . . .

112. Ibid., II.14, p. 189: *Sed et in prouincia Deirorum . . . baptizabat in fluuio Sualua, qui uicum Cataractam praeterfluit.* . . .

113. Ibid., II.16, p. 193: . . . *narrauit mihi presbyter et abbas quidam . . . rettulisse sibi quendam seniorem, baptizatum se fuisse die media a Paulino episcopo . . . in fluuio Treenta iuxta ciuitatem quae lingua Anglorum Tiouulfingacaestir uocatur.* . . .

114. Ibid., IV.16. Wallace-Hadrill (*Bede's Ecclesiastical History,* 156) points out this section as this chapter's only section not bearing on the conversion. For the first description of Wight, in the context of Vespasian's campaign, see Bede *EH* I.3.

115. Bede *EH* I.33, p. 115: *Fecit autem et monasterium non longe ab ipsa ciuitate ad orientum.* . . .

tle]."[116] Cedd, a man of God, came to the land of East Saxons and established new churches, "especially in the city called *Ythancaestir* in the Saxon tongue [Bradwell-on-Sea] and also in the place called Tilbury." Bede further specifies, "The former is on the river *Penta* [Blackwater] and the latter on the banks of the Thames."[117]

Bede also pays attention to the places of his characters' missionary and ecclesiastical activities. Relating how Augustine consecrated two bishops to preach in the province of the East Saxons and Kent, Bede tells us about the location and chief cities of these provinces.[118] Wilfrid went to preach to the kingdom of the South Saxons, "which stretches south and west from Kent as far as the land of the West Saxons and contains 7,000 hides."[119] Paulinus preached at the kingdom of Lindsey, "the first land on the south bank of the river Humber, bordering on the sea."[120]

Thus pointing out the locations, Bede maps out the spiritual progress of Christianity from one material place to another. For him, every physical place matters—whether a province accepting conversion, a newly founded monastery, or a bishopric receiving a new leader—because they all add to the spiritual achievements of the English church, demonstrating the work of divine providence. Because he believed that divine providence also directed the actions of the Anglo-Saxon kings, Bede included geographical information about their movements. His procedure in collecting details here resembles what he used elsewhere, in his reference work that listed bibilical places *[Nomina locorum]*; he gives the name of the place and its etymology and then describes the nature and political significance of the place.[121]

Places also matter because some of them can directly testify to the work of divine providence, providing a physical link between the divine and the human, which often corresponds, respectively, to the past and the present. The places where miracles happened possess miraculous power, which

---

116. Ibid., III.19, p. 271: *Erat autem monasterium siluarum et maris uicinitate amoenum, constructum in castro quodam quod lingua Anglorum Cnobheresburg, id est Urbs Cnobheri, uocatur.* ...

117. Ibid., III.22, pp. 283, 285: ... *maxime in ciuitate quae lingua Saxonum Ythancaestir appellatur, sed et in illa quae Tilaburg cognominatur; quorum prior locus est in ripa Pentae amnis, secundus in ripa Tamensis.* On other monasteries' localization see IV.3, pp. 336–37; IV.6, pp. 354–56.

118. Ibid., II.3.

119. Ibid., IV.13, p. 373: ... *quae post Cantuarios ad austrum et ad occidentem usque ad Occidentales Saxones pertingit, habens terram familiarum VII milium.* ...

120. Ibid., II.16, pp. 191, 193: *Praedicabat autem Paulinus uerbum etiam prouinciae Lindissi quae est prima ad meridianam Humbrae fluminis ripam, pertingens usque ad mare.* ...

121. See McClure, "Bede's Old Testament Kings," 94–95, with examples.

Bede often points out.[122] He reports that the walls of the church built by the first convert of Paulinus in the city of Lincoln "are still standing" and that "every year miracles of healing are performed in this place, for the benefit of those who seek them in faith."[123] Thus the blessing of conversion works through Paulinus, a saintly man in his own right, through his first convert, to the physical place transformed by the building of the church.

Bede reports that another special place, the one where Oswald, "the most Christian king of Northumbria," had set up a cross before a decisive battle and, upon a prayer, was granted victory by God, "is still shown today and is held in great veneration."[124] This place, according to Bede, was predestined for its role from the old days: its very name, Heavenfield, or Caelestis Campus in Latin, signified future heavenly events and miracles. Bede writes that, thus marked by providence and made sacred by Oswald's faith, the place has witnessed innumerable miracles of healing and has become even more sacred due to regular prayers of the local congregation and a church built there.[125] He notes that the place where Oswald was killed also performs miracles, healing sick men and beasts.[126]

Bede reports that the very soil from a holy place possesses miraculous power, working on all living things, from grass to humans. He relates how a Briton traveling near the place of Oswald's last battle "noticed that a certain patch of ground was greener and more beautiful than the rest of the field." The Briton immediately concluded—very wisely, in Bede's opinion—that the only cause of this must be "that some man holier than the rest of the army had perished there." Bede reports that the soil that the man took with him proved to cure sick people as well.[127]

---

122. On the Christian tradition of the veneration of holy places and their healing power see P. Brown, *The Cult of the Saints: Its Rise and Function in Latin Chrisitianity* (Chicago, 1981), esp. 86–126; B. Ward, *Miracles and the Medieval Mind: Theory, Record, and Event, 1000–1215* (Philadelphia, 1987).

123. Bede *EH* II.16, p. 193: . . . *parietes hactenus stare uidentur, et omnibus annis aliqua sanitatum miracula in eodem loco solent ad utilitatem eorum qui fideliter quaerunt ostendi.*

124. Ibid., III.9, p. 241: . . . *Osuald Christianissimus rex Nordanhymbrorum;* III.2, p. 215: *Ostenditur autem usque hodie et in magna ueneratione habetur locus ille, ubi uenturus ad hanc pugnam Osuald signum sanctae crucis erexit, ac flexis genibus Deum deprecatus est.* . . . On Oswald's image in historiography see Wallace-Hadrill, *Bede's Ecclesiastical History*, 88; in Bede's history, Hanning, *Vision of History*, 84–85.

125. Bede *EH* III.2.

126. Ibid., III.9: *Namque in loco ubi pro patria dimicans a paganis interfectus est, usque hodie sanitates infirmorum et hominum et pecorum celebrari non desinunt.*

127. Ibid., III.10, p. 245: . . . *et uidit unius loci spatium cetero campo uiridius ac uenustius, coepitque sagaci animo conicere, quod nulla esset alia causa insolitae illo in loco uiriditatis, nisi quia ibidem sanctior cetero exercitu uir aliquis fuisset interfectus. Tulit itaque de puluere terrae illius secum inligans in linteo, cogitans (quod futurum erat) quia ad medellam infirmantium idem puluis proficeret.* . . . Soil often figures in Bede's miracle stories: see III.11, on the soil on

Geographical descriptions in Bede's history served his goal—to record the progress of Christianity in Britain and, by implication, the work of divine providence that directed it. Bede, an exegete as well as a historian, placing his story in the context of several traditions, meant it to be understood on both literal and allegorical levels. Beginning his geographical introduction with a classical description, he joined the traditions of Roman geographers and Orosius and indicated his respect for these models. His idealized description of England and Ireland evoked the tradition of exegetical treatment of Paradise, thus creating the geographical setting that pointed at the level of "sacred history" behind the story of the English church. From his classical and idealized image in the introduction, Bede proceeded to a more precise and detailed geographical picture. He believed that divine providence, directing the progress of Christianity, manifested itself through places as well as events, and he mapped out this movement, trying to locate its steps: places where people received baptism, places where new churches and monasteries were founded, holy places that directly revealed the links between the human and the divine. The attention to places that Bede demonstrated in his history matches the interest in biblical locations that he revealed in his exegetical work.

Richer

Richer, a monk of the Abbey of St. Remigius at Reims, wrote his history in 991–98. As he says in the prologue, he was inspired by Gerbert, then archbishop of Reims, to describe the exploits of the Gauls, beginning close to his own time, with the reign of Charles the Simple (893–923).[128] At the end of his prologue, Richer declares that as his special goal, he wishes to relate everything in a clear and succinct manner. Immediately thereafter, he indicates the necessity of a geographical introduction: "I shall approach the beginning of my narrative after a brief description of the division of the world and the distribution of Gaul into parts, because I propose to describe the character and deeds of its peoples."[129] Thus Richer joins the

---

which the water, used for washing Oswald's bones, had been poured out; IV.3, on the soil from the grave of the saintly Bishop Chad; V.18, on the soil from the spot where Bishop Haedde died.

128. Richer, *Histoire de France,* ed. and trans. R. Latouche, 2 vols. (Paris, 1967), vol. 1, *Prologus,* p. 2 (hereafter cited as *Hist.;* page numbers in subsequent citations refer to this edition): *Gallorum congressibus in volumine regerendis imperii tui, pater sanctissime G., auctoritas seminarium dedit. . . . Cujus rei initium a vicino ducendum existimavi. . . .*

129. Ibid., vol. 1, *Prologus,* p. 4: *Ac totius exordium narrationis aggrediar, breviter facta orbis divisione Galliaque in partes distributa, eo quod eius populorum mores et actus describere propositum sit.*

tradition, well established in history writing, that required one to prepare a geographical scene before describing events. However, his previous declaration about brevity adds weight to his justification: even though he wants to be brief, he sees it fitting to spend some time on a geographical survey.

Following his list of geographical topics, Richer begins with the tripartite division of the earth, surrounded by the ocean. Next he gives the location of Asia, Africa, and Europe and the natural boundaries, rivers and mountains, between them.[130] This information is not directly relevant for the following historical account: most of the geographical names mentioned here will never appear again.[131] All this very general information goes back to the classical tradition, which Richer emphasizes by mentioning "cosmographers" in his first sentence. Like Orosius' and Jordanes' evocation of *maiores nostri,* Richer's reference to cosmographers places his survey in the classical context. His first sentence is so general that it is hard to tell if he had anybody specific in mind. Richer may have received the inspiration and basic information from Orosius; however, he did not follow Orosius' text word for word, and the order of the continents does not correspond to that of Orosius.[132]

After one paragraph of general exposition, Richer, true to his intention to be brief, narrows his focus down to Europe and, further, to Gaul: "Although each [part of the world] has its own subdivisions, I have decided to talk about only one part of Europe, which is called Gaul, from 'fairness,' because its people display characteristically fair faces." Richer borrows the etymology from Isidore, losing some of its clarity.[133] The fol-

---

130. Ibid., vol. 1, I.1, p. 6: *Orbis itaque plaga, quae mortalibus sese commodam praebet, a cosmographis trifariam dividi perhibetur, in Asiam videlicet, Africam et Europam. Quarum prior, a septemtrione per orientis regionem usque ad austrum, extrinsecus oceano disterminata, interius a Ripheis montibus usque ad terrae umbilicum, Thanai, Meothide, Mediterraneoque ab Europa distinguitur. . . .*

131. Except Africa and the Riphean Mountains. Defending himself against the accusations of hostile bishops, Gerbert appeals to the decisions of African and Toledan councils (ibid., vol. 2, IV.106, p. 326). Louis IV mentions the Riphean Mountains in no apparent connection to their geographical location (ibid., vol. 1, II.73, p. 248).

132. Robert Latouche suggests that Orosius had served as inspiration for Richer (ibid., vol. 1, pp. 1–2, n. 1. The order of the continents in Orosius, however, is Asia, Europe, and Africa (see Orosius *Hist.* I.2.1); in Richer the order is Asia, Africa, and Europe.

133. Richer *Hist.* vol. 1, I.1, p. 6: *Quarum singulae cum proprias habeant distributiones, Europae tamen partem unam quae Gallia a candore vocatur, eo quod candidioris speciei insigne eius oriundi praeferant, in suas diducere partes ratum duxi.* Cf. Isidore *Etym.* XIV.4.25: *Gallia a candore populi nuncupata est;* γάλα *enim Graece lac dicitur.*

lowing passage on the tripartite division of Gaul, containing the provinces of Belgica, Celtica, and Aquitanica, goes back to Julius Caesar.[134]

Richer's description of Gaul, with its boundaries and rivers, belongs to the classical picture of the world. Its tone—archaic, dry, and neutral—seems intentional in order to prepare the learned ground for his less impartial outline of the character of his countrymen. For his praise, Richer concentrates on the people rather than the land. The attention Richer devotes to this part—it occupies half of his introduction—testifies to its importance.

In Richer's description, all the people living in Gaul are brave, proud, and hot-tempered but intelligent and reasonable. To make his description sound more favorable, Richer quotes Jerome, who mentioned that Gaul, the only country that does not have monsters, has always been famous for its prudent and eloquent men.[135] The Belgae, just as in Caesar, are the bravest but also conduct their affairs with intelligence; the Celtae and Aquitani combine bravery and wisdom; the latter, although prone to enjoy their food, only do it to satisfy their nature. To support this, Richer turns to the authority of Sulpicius Severus, a bishop in Gaul and writer (d. 410), who defended his countrymen from the accusation of gluttony by explaining that "the voracious appetite among the Greeks comes from gluttony, among the Gauls from nature."[136]

In his geographical description, Richer does not name his sources, and he there paraphrases them rather than quoting them directly,[137] but in his description of the people of Gaul, he gives source names, using two vener-

---

134. Richer *Hist.* vol. 1, I.2, pp. 6–8. Cf. Julius Caesar *Commentarii rerum gestarum*, vol. 1, *Bellum Gallicum*, ed. W. Hering (Leipzig, 1987), I.1: *Gallia est omnis divisa in partes tres, quarum unam incolunt Belgae, aliam Aquitani, tertiam qui ipsorum lingua Celtae, nostra Galli appellantur.*

135. Richer *Hist.* vol. 1, I.3, p. 8: *Omnium ergo Galliarum populi innata audatia plurimum efferuntur, calumniarum impatientes. Si incitantur, cedibus exultant efferatique inclementius adoriuntur. Semel persuasum ac rationibus approbatum, vix refellere consuerunt. Unde et Hieronimus: Sola, inquit, Gallia monstra non habuit, sed viris prudentibus et eloquentissimis semper claruit.* Cf. Jerome, *Contra Vigilantium liber unus*, PL 23, 353–68, 355A.

136. Ibid., vol. 1, I.3, p. 10: *Celtae vero ac Aquitani consilio simul et audatia plurimi, rebus seditiosis commodi. Celtae tamen magis providi, Aquitani vero praecipites aguntur; plurimumque in ciborum rapiuntur appetitum. Quod sic est eis innatum ut praeter naturam non appetant. Hinc et Sulpicius: <u>Edacitas</u>, inquit, <u>in Graecis gula est, in Gallis natura</u>.* Cf. Caesar *Bellum Gallicum* I.3: *Horum omnium fortissimi sunt Belgae;* Sulpicius Severus, *Dialogi*, ed. C. Halm, CSEL 1 (Vienna, 1867), 160, I.8.

137. He treats them in much the same way as the main source of his history—the *Annals* of Flodoard. On his method see Richer *Hist.*, vol. 1, *Prologus*, p. 4: *ex quodam Flodoardi presbyteri Remensis libello me aliqua sumpsisse non abnuo, at non verba quidem eadem, sed alia pro aliis longe diversissimo orationis scemate disposuisse res ipsa evidentissime demonstrat.*

able Christian authorities to support his favorable description of the Gauls against the classical tradition. However, he still frames it in terms of this tradition rather than in terms of his own time: his emphasis on bravery echoes Caesar, and the accusation of gluttony also seems to be quite old—Sulpicius Severus' refutation refers to a reputation well established in the fourth and fifth centuries.

Richer, in his introduction, focuses on the classical image of Gaul and its inhabitants, which he seems to deliberately keep at a very general level, leaving elaborations and updates for the future. Caesar's tripartite division of Gaul provides Richer with the main frame that he consistently tries to follow. One of the corrections Richer made in the original version may be construed as his attempt to keep within Caesar's nomenclature of the provinces.[138] In the first version, Aquitanica had as one of its boundaries a province known by the name *Lugdunensis* (another, more recent name for Celtica). In his second version, Richer eliminated this name and replaced it by the indication of the Rhône and Saone Rivers as boundaries.[139]

Caesar's provincial names provide Richer with a general frame throughout his book.[140] He puts other, more recent divisions on this frame. Speaking of the Norman invasions, he introduces the province of Rouen as part of Gallia Celtica, and he introduces the province of Neustria as "part of Gallia Celtica that lies between the Seine and the Loire."[141] Later, he introduces Provence as "surrounded by the Rhône, the Alps, the sea, and Gothia" (northern Spain).[142] Richer puts contemporary ecclesiastical divisions on the same general frame.[143]

Throughout his history, Richer demonstrates a great interest in details,

---

138. The corrections Richer made in his manuscript are conveniently reproduced in both Latouche's and Waitz's edition. *(Richeri historiarum libri IIII*, ed. G. Waitz. 2nd ed. MGH SRG 51 [Hanover, 1877].)

139. Richer *Hist.* vol. 1, I.2, pp. 8–9, note e: *provinciae lugdunensi* is replaced by *Rhodano Ararique.* On the name Lugdunensis introduced under Augustus see M. Lugge, *Gallia und Francia im Mittelalter* (Bonn, 1960), 11.

140. For Belgica see Richer *Hist.* vol. 1, I.18, p. 154; II.19, p. 156; vol. 2, III.14.6, p. 14; III.81, p. 100; III.100, p. 128. For Celtica see vol. 1, I.4, p. 14; I.14, p. 36. For Aquitanica see vol. 1, I.6, p. 18; I.7, p. 20; II.39, p. 188; vol. 2, III.3, p. 10.

141. For Rouen see ibid., vol. 1, I.4, p. 12: . . . *Rhodomensem provinciam incolebant, quae est Celticae Galliae pars.* For Neustria see vol. 1, I.4, p. 14, text and notes: . . . *Galliae Celticae partem, quae Sequanae Ligerique fluviis interjacet, quae et Neustria nuncupatur, totam pene insectati sunt* (I have underlined the words added by Richer).

142. Ibid., vol. 1, I.7, p. 20: . . . *Provintia quoque, quae Rhodano et Alpibus marique ac Gothorum finibus circumquaque ambitur.* . . .

143. At the coronation of Charles the Simple at Reims, Belgica is represented by the bishops of Cologne, Treves, and Mainz; Celtica by the archbishop of Reims and the bishops of Laon, Chalons, and Terouanne; see ibid., vol. 1, I.12, p. 32.

including those concerning locations. His efforts to be geographically precise seem to be particularly directed at the military and administrative movements of his characters. He is interested in battles and in territorial changes that result from them.[144]

Flodoard's *Annals*, beginning in 919, form the main source for the first half of Richer's *History*, but Richer liberally supplies topographical details whenever he feels that Flodoard does not tell enough. When Flodoard, talking about the war between Louis IV and Otto the Great in 946, simply says that the German army devastated a number of places after crossing the Seine, Richer adds that the Germans advanced up to the Loire. Thus he adds another landmark and enhances the image of devastation.[145] Speaking of the same campaign, Richer makes Duke Hugh the Great retreat to Orleans, although, again, Flodoard keeps silent about it.[146]

In his second redaction, Richer seemed to pursue the same goal of increasing precision, as his numerous geographical corrections testify.[147] In the geographical introduction's section on Gaul, Richer adds information on the territory and boundaries of the three provinces.[148] He often

---

144. Körtum (*Richer von Saint-Remi*, 28–29) points out Richer's general interest in military affairs.

145. Richer *Hist.* vol. 1, I.58, p. 222 and note 1 on pp. 222–23: . . . *exercitus fluvium [Sequanam] transit. Dein terra recepti, incendiis praedisque vehementibus totam regionem usque Ligerim depopulati sunt.* Cf. Flodoard, *Annales*, ed. P. Lauer (Paris, 1905), 103: *Sicque trans Sequanam contendentes, loca quaeque praeter civitates gravibus atterunt depraedationibus terramque Nordmannorum peragrantes; loca plura devastant, indeque remeantes, regrediuntur in sua.*

146. Richer *Hist.* vol. 1, I.58, p. 222: *Dux vero Aurelianis sese receperat.* Cf. the quotation from Flodoard in n. 145 above. Other topographical additions include the descriptions of the devastation inflicted on the lands of Duke Hugh the Great by the army of Louis (Richer *Hist.* vol. 1, II.93, p. 282) and the reception of King Lothaire and his mother Gerberga by Duke Hugh the Great in Neustria in specific cities (vol. 2, III.3, p. 10).

147. These corrections should be studied in the context of all the changes that Richer made in his second redaction. However, this lies beyond the scope of the present inquiry, especially because, to my knowledge, no scholar has yet offered a comprehensive interpretation of the second redaction and its significance; only some aspects have been studied so far. See the summary and bibliography of the previous research in R. Latouche, "Un imitateur de Salluste au Xe siècle: L'historien Richer," *Annales de l'Université de Grenoble, nouvelle série, section Lettres-Droit* 6 (1929): 289–305, esp. 289–91.

148. At Richer *Hist.* vol. 1, I.2, p. 8, *Ab utroque vero latere, hinc quidem Alpibus Penninis, inde vero mari vallatur, cujus circumfusione insula Brittannica efficitur* is added on the margin, and the rather vague phrase *Celtica autem hinc Matrona inde Garumna abluitur* is changed to *Celtica autem a Matrona per longum in Garumnam distenditur.*

adds topographical indications missing in the first redaction, and he sometimes completely changes the itinerary of his characters.[149]

Because we do not know Richer's sources for most of these changes, it is hard to tell whether Richer "invented" his topographical details, as Robert Latouche, the most recent editor of Richer's text, claims.[150] Richer's interest in geographical details may have resulted from his following the rhetorical traditions that advised the description of places.[151] Whether or not his descriptions corresponded to any geographical reality, the important thing is that within his view of geographical space, he tried to introduce more details and thus increase precision and establish order.

Richer's ideas about topographical precision certainly do not always correspond to the "real world" (in our understanding of the term). In Flodoard's *Annals,* Conrad is the king of Cisalpine Gaul (meaning Burgundy). When Richer supplies the rare name *Genauni* for the subjects of Conrad and later provides them with a capital city—Besançon, "located in the Alps, with the river Doubs flowing by"—he tries to create his own geographical image, whose correspondence to reality does not seem to matter.[152] Richer apparently did not realize (or did not bother to realize) that Besançon was not located in the Alps. What he may have tried to do is to create a noncontradictory picture around a single classical term. He draws the name *Genauni* from Horace.[153] Then he supplies the people with a

---

149. He adds the destination of Charles the Simple on the margin at ibid., vol. 1, I.22, p. 55: *atque Remos devenit.* He makes the Normans invade the specific region of Burgundy rather than Gaul and adds the precise place of their victory at I.49, p. 98: *Galliarum* is changed into *Burgundie,* and *apud montem Calaum* is added on the margin. Other topographical indications are added at II.32, pp. 176–77, and II.93, p. 282. He makes Charles the Simple go to Saxonia instead of Belgica at I.14, p. 36: *[Karolus] in Saxoniam* [in the first redaction *Belgicam*] *secedit.* Similar changes occur at I.16, p. 38.

150. Latouche claims the same for details of chronology, numbers, and events added to Flodoard's account: see ibid., vol. 1: ix–x, 222–23 n. 1, 283 n. 3; Latouche, "Un imitateur," 297–98.

151. On the connections of Richer's history to rhetoric see Latouche, "Un imitateur"; Körtum, *Richer von Saint-Remi,* 93–112.

152. On the Genauni see Richer *Hist.* vol. 1, II.53, p. 214: *Nec minus et ab rege Genaunorum Conrado copias petit et accipit.* Cf. Flodoard, *Annales,* 102: *Qui [Otto], maximum colligens ex omnibus regnis suis exercitum, venit in Franciam, Conradum quoque secum habens, Cisalpinae Galliae regem.* On Besançon see Richer *Hist.* vol. 1, II. 98, p. 290: . . . *in urbem Vesontium, quae est metropolis Genaunorum, cui etiam in Alpibus sitae Aldis Dubis praeterfluit.* . . .

153. Horace, *Opera,* ed. D. R. Shackleton Bailey (Stuttgart, 1991), *Carm.* IV.14: *milite nam tuo / Drusus Genaunos, implacidum genus, / Breunosque velocis et arces / Alpibus impositas tremendis / deiecit acer plus vice simplici.* . . .

king, a capital city, and a count of this city. To do that, he consistently changes the information of Flodoard: Conrad, king of Burgundy, becomes the king of the Genauni; later Letoldus, a Burgundian count in Flodoard, turns into the count of Besançon, the capital of the Genauni. Besançon becomes a city in the Alps, because this is where the Genauni live, according to Horace. Richer consistently changes Flodoard's account—and geographical reality—to fit the information of his classical source.

Like many historians before him, Richer demonstrated tight connections between history and geography. He began with a geographical introduction where, following classical models, he created an image of Gaul that served as a frame for his historical account. Richer supplemented this classical frame by more recent details, thus demonstrating that the vision of geographical space in which his history developed differed from the classical one. However, his representation of space was still linked to the classical tradition. Trying to be precise and detailed in his indications of places, he constructed parts of his image with reference to the classical picture rather than the real world.

The four histories discussed in this chapter demonstrate a very close connection between geographical and historical matters. Orosius, seemingly the first historian who put a long geographical survey in the introduction to his work, justifying this by the goal of his history, set a pattern for other medieval history writings. In their respective geographical surveys, Jordanes, Bede, and, possibly, Richer directly used Orosius' material; all three got their inspiration from Orosius. How much every historian tells the reader about the earth in his introduction depends on his goals. In Orosius' universal history, the entire earth, observed from an elevated and remote point, provides the material for contemplating human sins and misery. Jordanes, Bede, and Richer focus on certain parts of the earth—Scandza, Britain, Gaul—important for their stories of nations. All four authors follow the classical geographical tradition and demonstrate a great regard for it. Their introductions either fully describe or evoke the timeless world of this tradition, Orosius and Richer keeping their image entirely classical, Jordanes and Bede introducing contemporary features. Although to different degrees, the four historians make their geography conform to the classical tradition: even when using information received elsewhere (for instance, from oral accounts), they put it in the context of traditional authority.

Both introductory surveys and geographical descriptions in the main text served and supported the overall goals and concepts of our four historians. In Orosius' and Bede's histories, biblical and exegetical models, shaping their historical concepts, also influenced their geographical descriptions. The earth not only provides the ground for contemplating human destiny but shares it in Orosius; in Bede, Britain, with its Paradise-like beginnings, moves toward the salvation of Christianity, whose progress Bede meticulously maps out. In Jordanes' and Richer's stories, the earth and particular locations on it provide the scene for historical action. Beginning with Scandza, places highlight Gothic history in Jordanes; Richer follows military and administrative movements of his characters, elaborating and updating the classical image of Gaul, but sometimes constructing his picture with regard to the inner logic of tradition rather than contemporary reality. Not only does geography serve history—it is inseparable from it. That historians freely used geographical writings and, like Orosius and Jordanes, established their own traditions, followed by geographical works, best demonstrates this deep connection and lack of strict disciplinary boundaries.

CHAPTER 4

Studying and Teaching Geography

Although not represented by an independent discipline, geographical knowledge occupied a place of its own in early medieval learning. That a number of texts were used as handbooks suggests that geography may have represented part of school education. These texts, revealing to us the contents and character of geographical studies at school, at the same time contribute to our understanding of early medieval education in general. Although newly discovered school texts keep coming to light, the story of geographical education in the Middle Ages still remains to be written. For my present focus, I have only chosen two traditions in geographical studies out of many—one begun by Isidore of Seville and another by Martianus Capella.

The influence exercised by Isidore on medieval culture and education in general and geographical knowledge in particular has made my first choice inevitable. It was within the frame of the Isidorean tradition that many important characteristics of geographical knowledge developed. Yet Isidore's contributions to geographical knowledge have so far attracted less attention than the influence of his encyclopedia on other fields of medieval learning.[1] At the same time, historical scholarship has accumulated a number of school texts, some recently discovered and edited, that would make a study of the Isidorean tradition possible.[2]

---

1. On Isidore's encyclopedia see the numerous works by Jacques Fontaine, especially the fundamental *Isidore de Seville et la culture classique,* which, however, does not consider geography; on the sources of Isidore's geographical books see H. Philipp, *Die historisch-geographischen Quellen in den Etymologiae des Isidorus von Sevilla* (Berlin, 1912); on the christianization of classical geography by Isidore see A. Melón, "La etapa isidoriana en la geografía medieval," *Arbor* 28 (1954): 456–67.

2. M. Zimmerman, "Le monde d'un catalan au Xe siècle: analyse d'une compilation isidorienne," in *Le metier d'historien au Moyen Age: Etudes sur l'historiographie médiévale,* ed. B. Guenée (Paris, 1977), 45–79; Gautier Dalché, *"Situs orbis terre";* a dialogue in Paris, BN lat. 4892, discovered by Gautier Dalché (*La "Descriptio mappe mundi,"* 97).

The influence of Martianus Capella's book VI, largely treating geography, has never become a special focus in scholarship, though much attention has been devoted to the medieval fate of Martianus' encyclopedia in general. The historians who have studied the manuscript transmission of Martianus' text have reconstructed the story of how, after being introduced into Carolingian schools in the second half of the ninth century, this encyclopedia came into wide circulation from England to the Alps. Scholars have also studied glosses and running commentaries that very often accompany Martianus' manuscripts and that testify to the intensive use of this text at school.[3] Commentaries to book VI, however, almost never become a focus of special attention.[4] Yet they deserve a special study. Written to clarify the text for students, these commentaries give us direct evidence of how early medieval people taught and studied geography at school and how they treated this particular text. In a wider context, these commentaries give us an insight into early medieval teaching techniques and ways of thinking.

The Isidorean Tradition

The way Isidore treated geographical knowledge in his *Etymologies*—an encyclopedia that incorporated classical sciences in the framework of Christian knowledge—defined methods and contents of later geographical studies at school. Developing a well-established late classical and early medieval tradition, Isidore made etymology both the main goal and the foundation of method.[5] He pointed this out, summarizing the contents of book XIII: "I have indicated certain causes of the sky and

---

3. On Martianus in medieval schools see C. Leonardi, "Nota introduttiva per un'indagine sulla fortuna di Marziano Capella nel Medioevo," *Bulletino dell'Istituto storico italiano* 67 (1955): 265–88, with bibliography; J. A. Willis, "Martianus Capella und die mittelalterliche Schulbildung," *Das Altertum* 19 (1973): 164–74; G. Glauche, *Schullektüre im Mittelalter: Entstehung und Wandlungen des Lektürekanons bis 1200 nach den Quellen dargestellt* (Munich, 1970), esp. 39–53. For the list of Martianus' manuscripts see Leonardi, "I codici"; J. Préaux, "Les manuscrits principaux du *De nuptiis Philologiae et Mercurii* de Martianus Capella," in *Lettres latins du Moyen Age et de la Renaissance* (Brussels, 1978), 76–128. On Martianus commentaries see Lutz, "Martianus Capella"; J. A. Willis, "Martianus Capella and His Early Commentators" (Ph.D. diss., University of London, 1952).

4. The exception is C. Leonardi, "Illustrazioni e glosse in un codice di Marziano Capella," *Archivio Paleografico Italiano*, n.s., 2–3, pt. 2 (1956–57): 39–60.

5. On etymology in late antiquity and the Middle Ages see R. Klinck, *Die lateinische Etymologie des Mittelalters* (Munich, 1970); M. Amsler, *Etymology and Grammatical Discourse in Late Antiquity and the Early Middle Ages* (Amsterdam and Philadelphia, 1989).

the location of the earth and the extent of the seas, so that the reader might quickly go through them and know, in a succinct form, their etymologies and causes."[6] For Isidore, to explain the etymology of a thing's name means to explain its cause; that is, to understand the nature and essence of a thing, one has to know the origin and meaning of the name: "For if you see where the name comes from, you will understand its force faster."[7] The two titles used for Isidore's work simultaneously and interchangeably in the Middle Ages, *Etymologies* and *Origins,* well reflect his idea.[8]

Isidore approached the world as a grammarian used to approach a text: he studied names and through their meaning arrived at the essence of things. In his account, things become almost indistinguishable from names, because the latter describe and supposedly reflect the properties of the former. For instance, the name of the created world, *mundus,* reflects its fundamental characteristic, which is to be in eternal motion *(in sempiterno motu).* Its Greek name, *kosmos,* reflects another characteristic, beauty, for "nothing that we see by our carnal eyes is more beautiful than the world."[9] The earth, along with the sky and all things created by God, constitutes part of the world, sharing its properties.

In Isidore's approach, there is no boundary between the word and the thing it signifies. He effortlessly switches from words to things and back to words. The word *terra* means both one of the four elements (air, fire, water, and earth) and a constituent part of the world, placed in its middle region and equidistant from any portion of the sky. Switching from thing to name, Isidore adds that *terra* also signifies, in the singular, the whole world *(totum orbem)* and, in the plural, its single parts. Various other names both signify and explain other aspects of the world—*orbis,* for

---

6. Isidore *Etym.* XIII: *In hoc vero libello quasi in quadam brevi tabella quasdam caeli causas situsque terrarum et maris spatia adnotavimus, ut in modico lector ea percurrat, et conpendiosa brevitate etymologias eorum causasque cognoscat.*

7. Ibid., I.29.2: *Nam dum videris unde ortum est nomen, citius vim ejus intellegis.*

8. On the title see Fontaine, *Isidore de Seville et la culture,* 1:11; on the equivalence of etymology and origin, Fontaine, "Isidore de Seville et la mutation," 532; on etymology as method in Isidore, J. Fontaine, "Cohérence et originalité de l'étymologie isidorienne," in *Homenaje a Eleuterio Elorduy, S.J.* (Bilbao, 1978), 113–44, reprint in J. Fontaine, *Tradition et actualité chez Isidore de Seville* (London, 1988), chap. 10, with bibliography.

9. Ibid., XIII.1–3: *Mundus est caelum et terra, mare et quae in eis opera Dei.... Mundus latine a philosophis dictus, quod in sempiterno motu sit, ut caelum, sol, luna, aer, maria.... Graeci vero nomen mundo de ornamento adcommodaverunt.... Appellatur enim apud eos κόσμος, quod significat ornamentum. Nihil enim mundo pulchrius oculis carnis aspicimus.*

instance, is so called in Latin because of its rounded form and because the ocean surrounds it.[10]

Thus beginning with the etymologies of the fundamental terms pertaining to the description of the earth, Isidore continues explaining etymologies of place-names. In his description of the provinces, his goal is to explain briefly "names and locations," as he says in the beginning of his account of Asia.[11] He derives his etymologies from both the Bible and classical sources. Thus, the name *Assyria* comes from Assur, the son of Shem, who first settled there after the flood; Canaan is named after the son of Ham or after the ten tribes expelled by the Jews; Scythia and Gothia are named after Magog, the son of Iaphet.[12] Many of Isidore's etymologies go back to antiquity and evoke mythological associations: for example, Europe was named after the daughter of the king of Libya, abducted by Jupiter; in its turn, Libya, that is, Africa, received its name from the king's mother.[13]

Since explaining names was Isidore's ultimate goal, he chose the logical arrangement of subjects by categories in his description of the world, thus abandoning the traditional order of classical geographical descriptions that followed an imaginary journey along the coastal line. According to his logical principle, he put places and peoples, usually treated together in classical geography, in separate books; describing the world, he treated his subjects following the order of the four elements (fire, air, water, and earth), thus separating seas and rivers from the dry land and its regions.[14]

---

10. Ibid., XIV.1.1: *Terra est in media mundi regione posita . . . quae singulari numero totum orbem significat, plurali vero singulas partes. Cuius nomina diversa dat ratio; nam terra dicta a superiore parte, qua teritur; humus ab inferiori vel humida terra . . . tellus autem, quia fructus eius tollimus . . .* ; XIV.2.1: *Orbis a rotunditate circuli dictus, quia sicut rota est. . . .*

11. Ibid., XIV.3.1: *Habet autem [Asia] provincias multas et regiones, quarum breviter nomina et situs expediam. . . .*

12. Ibid., XIV.3.10: *Assyria vocata ab Assur filio Sem, qui eam regionem post diluvium primus incoluit;* XIV.3.20: *Haec [Iudaea] prius Chanaan dicta a filio Cham, sive a decem Chananaeorum gentibus, quibus expulsis eandem terram Iudaei possiderunt;* XIV.3.31: *Scythia sicut et Gothia a Magog filio Iaphet fertur cognominata.*

13. Ibid., XIV.4.1: *Europa quippe Agenoris regis Libyae filia fuit, quam Iovis ab Africa raptam Cretam advexit, et partem tertiam orbis ex eius nomine appellavit. Iste est autem Agenor Libyae filius, ex qua et Libya, id est Africa, fertur cognominata. . . .* Philipp, in his study of Isidore's sources (*Die historisch-geographischen Quellen,* 106–7) lists several parallels (Festus, Hyginus, Servius), but since Isidore's text precisely corresponds to none of them, it is hard to establish his immediate source.

14. On Isidore's theory of the four elements and its sources see Fontaine, *Isidore de Seville et la culture,* 2:654–60.

As a result, there is no unified image of geographical space in Isidore's *Etymologies;* instead, his logical arrangement yields several different images, organized around nations, seas and rivers, and regions.

The account of nations, preceding the account of the earth, becomes entirely separated from it. Isidore discusses the names and locations of peoples in book IX, which is devoted to miscellaneous human institutions—"languages, peoples, kingdoms, warfare, citizens, and degrees of kinship."[15] The account of nations and the account of the earth belong to different fields of inquiry: the former deals with names and divisions that already exist, the latter with physical things that one has to distinguish by giving them names and thereby explaining them. For Isidore, the division of languages precedes and forms the foundation of the division into nations; the latter, in its turn, produces the divisions of the earth: "We have first treated the languages and then the nations, because the nations originate from languages, not languages from nations. . . . And the earth is divided by nations. . . ."[16] Thus Isidore divides the earth by nations, classifying them by their ancestors, the biblical sons of Noah (Iaphet, Shem, and Ham), and subordinating geographical divisions to this logical arrangement.

The description of waters in book XIII forms Isidore's second image of geographical space, split into categories, such as seas, gulfs, lakes, and rivers. Beginning with the ocean, so called because "it goes around the world like a circle," Isidore names its seas, moving from north to south.[17] He then devotes a section to the Mediterranean Sea, which divides the three continents: Europe, Africa, and Asia. This section is the most detailed description in book XIII, organized around the area that Isidore refers to as the gulfs of the Mediterranean: Iberian, Gallic, Tyrrhenian, and so on. Isidore highlights the names of these gulfs and their location, sometimes supplying, very briefly, names of towns, islands, or historical figures. In the rest of book XIII, Isidore focuses on explaining the names of his broad categories—gulfs or lakes—rather than on reporting locations of single objects. He only mentions the latter to exemplify the meaning of names.[18]

---

15. Isidore *Etym.* IX: *De linguis, gentibus, regnis, militia, civibus, affinitatibus.*

16. Ibid., IX.I.14: *Ideo autem prius de linguis, ac deinde de gentibus posuimus, quia ex linguis gentes, non ex gentibus linguae exortae sunt;* IX.2.2: *Gentes autem a quibus divisa est terra.* . . .

17. Ibid., XIII.15.1: *Oceanum Graeci et Latini ideo nominant eo quod in circuli modum ambiat orbem.*

18. For instance, ibid., XIII.17.1: *Sinus dicuntur maiores recessus maris, ut in mari Magno Ionius, in Oceano Caspius, Indicus, Persicus, Arabicus, qui et mare Rubrum, qui Oceano adscribitur.*

Rather than forming a single geographical image, Isidore represents his "waterworld" as a combination of terms and names.

Finally, in book XIV Isidore describes the provinces of the earth. In this third image, the earth of antiquity coexists with the earth of the Bible.[19] The Bible provides the basic framework—Isidore's description begins with Paradise and ends with Hell. However, the description of the regions follows the classical tripartite division into continents, which Isidore does not connect to the settlement of Noah's sons as he did in his account of nations. Filling the Christian frame with classical contents, Isidore derives his geographical information from Orosius, Pliny, and Solinus.[20]

Constructing his image with reference to the Bible and classical authorities, Isidore shows no interest in the contemporary world, with its new peoples and political boundaries. In his image, Saracens still live in the Arabian Desert, and the most recent information about Gothia is that it is called after the biblical Magog and is located near Dacia.[21] And yet Isidore records more recent events elsewhere in the *Etymologies*—in his brief outline of history, he mentions the conversion of the Goths to Christianity and "the most religious King Sisebut," who ruled Visigothic Spain in Isidore's time.[22] Confining these facts to the historical chapter, Isidore seems to reveal an important assumption—that geographical knowledge, the way he understood it, did not require contemporary information.

Isidore's synthesis of Christian and classical knowledge, including geography, proved to be very influential in the Middle Ages, as numerous manuscripts testify.[23] His *Etymologies,* widely copied and studied at schools, transmitted his ideas about geographical knowledge—the etymological method, the logical arrangement of subjects, and the combination of Christian and classical pictures of the world.

Both Isidore's etymological method and Isidore's material were extensively used in a school commentary on the two first books of Orosius' *History against the Pagans*. Preserved in a ninth-century manuscript, this commentary may have been composed by an Irish master in Reims,

---

19. On Isidore's Christianization of classical geography see Melón, "La etapa isidoriana."

20. On Isidore's sources in detail see Philipp, *Die historisch-geographischen Quellen.*

21. Isidore *Etym.* IX.2.57: *Saraceni dicti, vel quia ex Sarra genitos se praedicent, vel sicut gentiles aiunt, quod ex origine Syrorum sint, quasi Syriginae. Hi peramplam habitant solitudinem;* XIV.3.31: *Scythia sicut et Gothia a Magog filio Iaphet fertur cognominata* (cf. IX.2.27 and IX.2.89); XIV.4.3: *... post hanc [Alaniam] Dacia, ubi est Gothia ...*

22. Ibid., V.39.41–42: *Gothi catholici efficiuntur. . . . Huius quinto et quarto religiosissimi principis Sisebuti Iudaei in Hispania Christiani efficiuntur.*

23. For the list of Isidore's manuscripts see Fernandez Caton, *Las Etimologias.*

between the seventh and the ninth centuries.[24] The commentator consistently supplemented Orosius' geographical account by etymologies, both Christian and classical, mainly derived from Isidore. Whereas Orosius simply says "the earth" *(orbis terrarum)*, the commentator, following Isidore, gives the reader all the terms used to designate the earth and its parts: *terra* comes from *teritur* (is trod upon); *humus* (soil) comes from the humidity of the earth; *orbis* is so called because it resembles a circle.[25] Throughout the commentary, etymology serves as the main method of explanation; almost every gloss begins with it, and all other information is attached to it. Whereas Orosius simply names the three continents (Asia, Europe, and Africa), the commentator adds etymologies, drawing on Isidore and other sources. Thus, Europe has received its name from a Libyan princess abducted by Jupiter; Africa has been so called because its name resembles the word *aprica* (open, exposed) and the whole region is exposed to the sun.[26] In this commentary, Orosius' essentially classical geographical image receives an entirely different etymological dimension. Interpreting Orosius' text and uncovering its meaning, the commentator supplements Orosius' timeless and formal image of the world by one no less timeless and formal—the image consisting of and related to words.

Another work in the same tradition, an Isidorean compilation composed in the tenth century at Ripoll, carried Isidore's etymological method almost to its logical extreme. Consisting of a number of treatises on the liberal arts, this compilation also included an abbreviated version of Isidore's book XIV, under the name of *Book on the Arrangement of the World*.[27] This text considerably reduces and simplifies Isidore, only retaining 20 percent of his classical quotations from nongeographical sources and about one-third of his place-names, and excluding more complicated etymologies.[28] One of a very few additions of his own that the Ripoll compiler

---

24. Vatican, Reg. lat. 1650, s. IX 2/2, Soissons or Rheims. See Szerwiniack, "Un commentaire," 6–7, 9 (on the educational purpose), 30–34 (on the date and place of composition).

25. Szerwiniack, "Un commentaire," pp. 47–48: *Terra dicta a superiore parte qua teritur; humus ab inferiori uel umida terra, ut sub mari . . . Orbis de rotunditate circuli dictus, quia sicut rota est. . . .* Cf. Isidore *Etym.* XIV.1.1, XIV.2.1.

26. Szerwiniack, "Un commentaire," p. 49: *Eurupa [sic] quippe Agenoris regis Libiae filia fuit, quam Iouis ab Africa raptam, Cretae aduexit, et partem tertiam eius nomine appellauit;* on Africa: *Aliter ut Isidorus dicit: Africa est apellata quasi aprica, quod sit aperta caelo et soli et sine horrori frigoris.* Cf. Isidore *Etym.* XIV.4.1, XIV.5.2.

27. *Liber de mundi institucione,* Barcelona, Archivo de la Corona de Aragon, Ripoll 74, s. X 2/2, Ripoll, fols. 94rb–98vb, edited in Zimmerman, "Le monde" (page numbers in subsequent citations of this text refer to Zimmerman's edition).

28. Ibid., 53–54, 57–58, 68–69.

inserted between Isidore's lines demonstrates the compiler's etymological emphasis and, by implication, his main goal. After summarizing Isidore's statement on the importance of place-names, the compiler explains where they can come from: "A name of a province is derived from the name of a progenitor, or the name of a river, or the name of a king, or the name of a city, or the merit of the land."[29] Thus very aptly summarizing Isidore's method, the Ripoll compiler gives the reader a methodological key to understanding and, maybe, composing etymologies.

Words and names seem to be this compiler's main emphasis, even more than Isidore's. If Isidore, in his geographical book, wanted briefly to give names and locations of regions and provinces,[30] the Ripoll compiler only gives place-names, very rarely specifying where the places are. Whereas Isidore, even studying words, had essentially retained the classical geographical image, the Ripoll compiler almost entirely composed his image of words. His "arrangement of the world," entirely verbal and formal, is even farther removed from reality than Isidore's. And yet, as in Isidore's *Etymologies,* other parts of the same compilation contain contemporary realities.[31] The Ripoll compiler, like Isidore, seems to have thought that a picture of the world did not need to be contemporary.

Another of Isidore's contributions, his logical arrangement of subjects, also influenced a number of geographical texts. A school text based on Isidore and Orosius, an eighth- or ninth-century dialogue from a Spanish cultural circle,[32] makes Isidore's logical method particularly obvious. It lists the provinces, mountains, and rivers of each continent in an order roughly corresponding to that of Isidore and concludes with a brief account of the three continents and their location and a list of seas. This text entirely omits descriptions and reduces its geographical image to that of a logically arranged catalogue of names, carrying Isidore's formalization even further.

Both Isidore's etymological method and his logical arrangement influenced the ninth-century geographical treatise by Anonymus Leiden-

---

29. *Liber de mundi institucione,* p. 76: *Nomen provinciarum aut dirivantur ex nomine auctorum aut ex nomine fluminum aut ex nomine principum aut ex nomine civitatum aut ex merito terrarum.*

30. Isidore *Etym.* XIV.3.1: *Habet autem [Asia] provincias multas et regiones, quarum breviter nomina et situs expediam. . . .*

31. See Zimmerman, "Le monde," 72–73.

32. Paris, BN lat. 4892, edited in Gautier Dalché, *La "Descriptio mappe mundi,"* on the date and origin see p. 97.

sis, *On the Location of the Earth*.[33] Indicating his intentions in the prologue, Anonymus Leidensis echoes Isidore's wish to treat etymologies and causes of the created world, and he defines the theme of his first book as a brief account "of both seas and the causes of their names as well as islands located in them."[34] Following Isidore's method and borrowing his material, Anonymus Leidensis begins the account of the ocean and seas with the explanation of their names and often uses Isidore's etymologies in the account of the regions. Once again, we learn that the ocean is so called because it encircles the earth; that Africa derives its name from its openness to the sun; that the name of Egypt goes back to the brother of Danaus.[35] Etymologies provide the starting point in the geographical description; all other material—location, peoples, curiosities—follows them.

Isidore's logical order influenced the way Anonymus Leidensis presented his material, devoting his first book to seas and islands and his second to the description of regions. Isidore also provided much of the general structure within book I, where Anonymus Leidensis uses quotations from the *Etymologies* to organize his account of the seas.[36]

Everything that Isidore had contributed to geographical knowledge—both his methods and his material, copied word for word—was used by Hrabanus Maurus in his encyclopedia *On the Natures of Things*. Despite his copious borrowings, Hrabanus pursued his own goals and made his own contributions to geographical knowledge.[37]

Fully using Isidore's etymological approach, Hrabanus added to it the allegorical method borrowed from biblical exegesis. Hrabanus, as he declared in his prologue, wrote his book in order to explain "the natures of

---

33. Anonymus Leidensis *De situ orbis*.

34. Anonymus Leidensis *De situ orbis, Proemium* 3: . . . *totumque opusculum in duobus libellis distinxi, ita ut primus utriusque maris vocabulorumque eorum causas nec non insularum quae in eis sunt sitae, quamvis breviter digestas contineret*. . . .

35. Ibid., I.2.1: *Oceanum et Greci et Latini ideo nominant, eo quod in circuli modo ambiat orbem* . . . (= Isidore *Etym.* XIII.15.1); I.4.3: *Africam autem nominatam quidam exinde existimant, quasi apricam, quod sit aperta caelo vel soli* . . . (= Isidore *Etym.* XIV.5.1); I.4.4: *Aegyptus autem ab Aegypto Danai fratre* (= Isidore *Etym.* XIV.3.27).

36. See Quadri, introduction to Anonymus Leidensis, *De situ orbis libri duo,* xxii; Gautier Dalché, "Tradition," 131; N. Lozovsky, "Carolingian Geographical Tradition: Was It Geography?" *Early Medieval Europe* 5 (1996): 25–43.

37. The accusation of plagiarism often directed at Hrabanus by historians makes no sense in view of what we know about medieval methods of composition shared by Hrabanus. See the discussion with bibliography in T. Burrows, "Holy Information: A New Look at Raban Maur's *De Naturis Rerum,*" *Parergon* 5 (1987): 28–27, and W. Schipper, "Rabanus Maurus *De rerum naturis:* A Provisional Check List of Manuscripts," *Manuscripta* 33 (1989): 109–18.

things and properties of words and also . . . the mystical signification of things."[38] Applying exegetical methods to the visible world, Hrabanus wished to describe and explain "things" both literally and allegorically, or, as he sometimes put it, "historically" and "mystically."[39] Approaching the account of the created world, Hrabanus once again emphasized the necessity of these interpretations: "we shall first indicate the things themselves; then briefly treat their significance, so that, beginning with the world, we would then discuss its parts."[40] Hrabanus, quite in the Isidorean tradition, treats the created world as a text, but whereas Isidore focuses on its grammar, Hrabanus tries to uncover its symbolical meaning.[41]

Essentially repeating Isidore's account of the world, using the same etymological method and logical arrangement, Hrabanus added his own emphasis. In his encyclopedia, seas and lands, rivers and mountains are simultaneously created things, when he describes them historically (that is, gives their location and sometimes dimensions, population, etc.), and the signs of another, spiritual reality, when he interprets them allegorically. So, faithfully reproducing Isidore's etymologies of the word *terra* (earth), Hrabanus adds to them the "mystical" explanation: it can signify the heavenly motherland, or the flesh of God the Savior, or the Mother of God.[42] He proceeds in the same way, explaining, for instance, that the tripartite division of the earth signifies the Trinity; that India designates the church; that Assur, who in Isidore gave his name to Assyria, means the devil.[43]

The new allegorical dimension that Hrabanus adds to Isidore's verbal picture carries geographical knowledge one more step away from geographical reality as we understand it. Literal treatment, that is, the traditional geographical description, far from being a goal, becomes a means of uncovering another reality behind it. As a result, Hrabanus' image of the

---

38. Hrabanus *De rerum naturis, Praefatio ad Ludovicum regem invictissimum Franciae,* 9B: *Sunt enim in eo plura exposita de rerum naturis, et verborum proprietatibus, nec non etiam de mystica rerum significatione.*

39. Ibid., *Praefatio altera ad Hemmonem episcopum,* 12D–13A: *. . . in quo haberes scriptum non solum de rerum naturis et verborum proprietatibus, sed etiam de mystica earumdem rerum significatione ut continuatim positam invenires historicam et mysticam singularum expositionem.*

40. Ibid., IX, *Prologus,* 259A: *. . . res ipsas primum notamus: deinde significationem earum breviter disseremus, ita ut a mundo incipientes, partes ejus subsequenter disserendo ponamus.*

41. On medieval allegory, its methods, and its significance see F. Ohly, "Vom geistigen Sinn des Wortes im Mittelalter," in *Schriften zur mittelalterlichen Bedeutungsforschung* (Darmstadt, 1977), 1–31; C. Meier, "Das Problem der Qualitätenallegorese," *Frühmittelalterliche Studien* 8 (1974): 385–435.

42. Hrabanus *De rerum naturis* XII.1.331B–D.

43. Ibid., XII.4.335 A–D.

world appears as formal as that of Isidore and even more abstract. Hrabanus displays as little interest in contemporary realities as did Isidore; his image, in its literal component going back to Isidore and Isidore's Roman sources, becomes even more archaic by the ninth century.

True to his intention to help people reading the Scriptures, Hrabanus Maurus Christianized the geographical picture even more consistently than Isidore had done. Structuring his account by beginning with Paradise and ending with Hell, at the center of the earth Hrabanus places Jerusalem. To do this, Hrabanus changes the text of Isidore, who simply says that "Jerusalem is in the middle of Judea, as if the navel of the whole region."[44] Hrabanus adds, "and the whole earth."[45] Describing Palestine, Hrabanus more often associates the places with biblical events, adding this information to that of Isidore. When Isidore talks about Mesopotamia among other regions in Asia, he only gives the Greek etymology of its name and its boundaries: "Mesopotamia has a Greek etymology because it is surrounded by two rivers. . . ." Hrabanus, faithfully copying this, adds that Jacob had taken his wife there and proceeds with the allegorical interpretation.[46] Isidore, mentioning Babylonia, only reports the etymology of its name; Hrabanus, in addition to this, reminds his readers that its king Nebuchadnezzar had burned the Temple.[47] As a result, Hrabanus' image of the world becomes much more centered on the Bible than Isidore's.

Transmitted and developed in school and reference texts, the Isidorean tradition shaped the development of early medieval geographical knowledge in several ways. Isidore and his followers produced an image of the world, combining Christian and classical features. This image, organized by categories, was formal and removed from contemporary reality. Treat-

---

44. Isidore *Etym.* XIV.3.21: *In medio autem Iudaeae civitas Hierosolyma est, quasi umbilicus regionis totius.*

45. Hrabanus *De rerum naturis* XII.4.339C: *In medio autem Judaeae civitas Hierosolyma est, quasi umbilicus regionis et totius terrae.* . . .

46. Isidore *Etym.* XIV.3.13: *Mesopotamia Graecam etymologiam possidet, quod duobus fluviis ambiatur; nam ab oriente Tigrim habet, ab occiduo Euphraten. Incipit autem a septentrione inter montem Taurum et Caucasum; cuius a meridie sequitur Babylonia, deinde Chaldaea, novissime Arabia Εὐδαίμων.* Cf. Hrabanus *De rerum naturis* XII.4.337A: *Mesopotamia Graecam etymologiam possidet, quod duobus fluviis ambiatur. Nam ab oriente Tigrim habet, ab occiduo Euphraten. Incipit autem a Septentrione inter montem Taurum et Caucasum, cujus a meridie sequitur Babylonia, deinde Chaldaea, novissime Arabia Eudaemon. Mesopotamia, quae interpretatur elevata, unde Jacob ducit uxorem.* . . .

47. Isidore *Etym.* XIV.3.14: *Babyloniae regionis caput Babylon urbs est, a qua et nuncupata, tam nobilis ut Chaldaea et Assyria et Mesopotamia in eius nomen aliquando transierint.* Cf. Hrabanus *De rerum naturis* XII.4.337C: *In qua [Babylonia] regnavit Nabuchodonosor. . . . Ipse enim vastavit civitatem Dei, et templum incendit.* . . .

ing the world as a text, the Isidorean tradition used etymology as a key to understanding and describing it. While very influential, the Isidorean tradition was by no means the only one in early medieval geographical studies. To reconstruct one more facet of these studies, I shall now consider how early medieval scholars approached another tradition—that of Martianus Capella.

Martianus' Geography at School

In this section I shall focus on commentaries to book VI of Martianus Capella's encyclopedia, transmitted in ninth- and tenth-century manuscripts. Scholars have identified several versions of the Martianus commentary, attributing some of them to various ninth-century scholars, such as Martin of Laon, John Scottus Eriugena, and Remigius of Auxerre.[48] The problem of attribution lies beyond the scope of the present work, and the limited manuscript evidence that I use would not allow me to consider it. Moreover, an attribution based on the modern concept of authorship seems problematic when we deal with glosses—interlinear and marginal notes, ranging in length from one to several dozen words. These medieval teachers' or students' notes are as hard to attribute to a single author as are modern lecture notes or comments on the margin of a textbook. Attribution may become less problematic when we deal with glosses transmitted as a running commentary (as is the case with Remigius' commentary,

---

48. The only manuscript of Martin of Laon (Paris, BN lat. 12960, s. IX, Corbie) where fragments survive as a running commentary was edited by Cora Lutz and attributed to Dunchad in Dunchad, *Glossae in Martianum,* ed. C. Lutz (Lancaster, 1944); later Jean Préaux suggested the attribution to Martin of Laon, in J. Préaux, "Le commentaire de Martin de Laon sur l'oeuvre de Martianus Capella," *Latomus* 12 (1953): 437–59; this attribution was questioned in J. Contreni, "Three Carolingian Texts Attributed to Laon," *Studi medievali,* 3d ser., 17 (1976): 802–13, and *The Cathedral School of Laon: Its Manuscripts and Masters* (Munich, 1978), 114. John Scottus' commentary was edited by Cora Lutz from Paris, BN lat. 12960, in *Iohannis Scotti Annotationes in Marcianum.* Later another version of this commentary was discovered in MS Oxford, Bodleian, Auct. T. 2.19, s. IXex–Xin, [Metz?]; see L. Labowsky, "A New Version of Scotus Eriugena's Commentary on Martianus Capella," *Mediaeval and Renaissance Studies* 1 (1941–43): 187–93. The attribution to John Scottus has also been questioned: see G. Schrimpf, "Zur Frage der Authentizität unserer Texte von Johannes Scottus' *Annotationes in Martianum,*" in *The Mind of Eriugena: Papers of a Colloquium, Dublin, 14–18 July 1970,* ed. J. J. O'Meara and L. Bieler (Dublin, 1973), 49–58, and M. Herren, "The Commentary on Martianus Attributed to John Scottus: Its Hiberno-Latin Background," in *Jean Scot écrivain: Acts du IVe colloque international, Montréal, 28 aôute–2 septembre 1983,* ed. G.-H. Allard (Montreal and Paris, 1986), 265–86, esp. 267–69. Remigius of Auxerre's commentary is edited by Cora Lutz in *Remigii Autissiodorensis Commentum in Martianum Capellam.*

edited by Cora Lutz). Still, variations, sometimes considerable, may challenge the whole concept of attribution.[49]

Even though attribution seems problematic, the commentaries to Martianus Capella do fall into distinct groups identified by Claudio Leonardi and Cora Lutz. For the sake of convenience, I shall keep the names of the groups that they use, referring to the different traditions, for instance, as "Martin's" or "John Scottus'" commentaries.

Despite much work done on Martianus commentaries, glosses to book VI have never become a focus of special attention. Neither Leonardi's nor Lutz's lists give information concerning the number of glosses to separate books, and in some cases glosses to book VI are very few. I shall focus on the manuscripts that contain glosses or running commentaries to book VI. Although the absence of glosses in a manuscript does not necessarily mean that book VI was not taught or studied at that center, their presence certainly demonstrates a special attention to Martianus' text.

Manuscripts—Physical Evidence

The ninth- and tenth-century manuscripts testify to a considerable popularity of Martianus Capella's book VI in Continental and English centers of study. Most manuscripts contain Martianus' text accompanied by interlinear and marginal glosses; some transmit separate running commentaries. Tables 1–4 provide a summary of information about the manuscripts containing commentaries to book VI.

The manuscript evidence points to Reims as the center that may have produced the earliest version of a Martianus commentary.[50] Auxerre and Reims seem to have paid particular attention to Martianus' book VI. The manuscripts with the most numerous glosses, belonging to several groups, come from there.[51] Apart from the attention to Martianus' book VI, these

---

49. For instance, the difference between the text of John Scottus' commentary in Paris and Oxford manuscripts led some scholars to postulate two separate versions written by John Scottus: see Schrimpf, "Zur Frage"; Herren, "Commentary," 269. On the attribution problems see also Marenbon, *From the Circle,* 116–19; A. M. White, "Glosses Composed before the Twelfth Century in Manuscripts of Macrobius' Commentary on Cicero's Somnium Scipionis" (Ph.D. diss., Oxford University, 1981), 33–34.

50. Jean Préaux ("Le commentaire," 439–41) suggests that the glosses in Leiden, B.P.L. 88, present the earliest version of the commentary that he attributed to Martin of Laon. Bernhard Bischoff identified the hand of the main text as belonging to Reims: Bischoff is cited in J. Préaux, "Deux manuscrits gantois de Martianus Capella," *Scriptorium* 13 (1959): 15–21; on Bischoff's identification see ibid., 19.

51. The manuscripts from Reims may include "Martin," Leiden, B.P.L. 88, and Vatican, Reg. lat. 1987; possibly Anonymus Cantabriensis, Cambridge, C.C.C. 330, pt. II. Those from Auxerre may include "Martin," Leiden, Voss. lat. F. 48, and Vatican, Reg. lat. 1535 (few

TABLE 1. "Martin"

MS	Date	Origin	Glosses to Book VI	Diagrams
Leiden, B.P.L. 88	IX 2/2	Reims	2–3 hands; contemporary or slightly later	
Leiden, Voss. lat. F. 48	IX (847?)	[Auxerre?]	1 hand contemporary; 2 hands slightly later	
Besançon, Bibl. Mun. 594	IX 2/2	[St Oyan?]	Same hand	Sundial circles
Florence, San Marco 190	Xex–XIin	France	Same hand	Sundial circles; sundial view; map
Vatican Reg. lat. 1987	IX	[Reims?]	Slightly later	
Vatican, Reg. lat. 1535	IX	[Auxerre?]	Very few; same or contemporary	
Leiden, B.P.L. 87	IX		Very few	
St. Petersburg, Publ. Bib., Clas. lat. F.V. 10	X	Corbie	Contemporary	
Paris, BN lat. 8670	IX	Corbie	Very few; contemporary	

centers appear to demonstrate a special interest in geographical studies in the second half of the ninth century by copying other geographical materials. The only manuscript of Anonymus Leidensis' tract *On the Location of the Earth* (Leiden, Voss. lat. F. 113, pt. II) was written at Auxerre in the

---

glosses to book VI); Remigius, version B, Paris, BN lat. 8671; unidentified, Paris, BN lat. 8671. On geographical studies at Auxerre see P. Gautier Dalché, "La géographie descriptive à Saint-Germain d'Auxerre (milieu IXe–début Xe siècle)", in Saint-Germain d'Auxerre: Intellectuels et artistes dans l'Europe carolingienne, IX–XI siècles (Auxerre, 1990), 270–76. I thank Anna Somfai for our discussions of manuscripts and glosses, which led to these conclusions.

TABLE 2. Remigius, version B

MS	Date	Origin	Glosses to Book VI	Diagrams
Paris, BN lat. 8674	X	Auxerre	Running commentary	
Paris, BN NAL 340	X	Cluny	Running commentary	Sundial circles; describes them

Note: This version resembles Remigius' commentary edited by Lutz (version A) but presents a separate recension with minor differences in books I–V and considerable differences in books VI–IX; it may have resulted from notes taken by one of Remigius' students—see Lutz, "Martianus," 372.

TABLE 3. Anonymous Cantabriensis

MS	Date	Origin	Glosses to Book VI	Diagrams
Cambridge, C.C.C. 153	1. IXex–Xin	Wales	Interlinear contemporary insular	
	2. X 3/4	England	Running commentary	
Cambridge, C.C.C. 330	1. XI–XII	England	Interlinear contemporary	
	2. IXex	[Reims?]	Running commentary	
Paris, BN lat. 8669	IX	[Soissons?]	Interlinear and marginal, 1–2 hands, contemporary	Unfinished sundial circles

second half of the ninth century; the author, who knew and extensively used Martianus Capella, may have belonged to the group of masters or students who studied Martianus' book VI.[52] The anonymous author also used Pomponius Mela's *Chorography;* a ninth-century manuscript containing this and other geographical works (Vatican, Vat. lat. 4929) was

---

52. Quadri, introduction to Anonymus Leidensis, *De situ orbis libri duo,* lvi; Gautier Dalché, "Tradition," 138.

TABLE 4. Unidentified

MS	Date	Origin	Glosses to Book VI	Diagrams
Cologne, Dombibl. 193	X–XI		Few; contemporary or slightly later	
Leiden, B.P.L. 36	IX	[Lorsch?]	Contemporary	Sundial circles
Paris, BN lat. 8671	IX 2/3	[Auxerre?]	Contemporary; similar	
Vienna, Nationalbibl. 177	X		Same hand	Describes sundial circles

read and corrected by Heiric, a disciple of John Scottus who worked at Auxerre in the 860s and 870s.[53] Paris, BN lat. 4806, produced at Reims in the third quarter of the ninth century, contained Dicuil's tract *On the Measurement of the Earth* and other geographical texts.[54] Glosses on Orosius' geographical chapter, transmitted in manuscript Vatican, Reg. lat. 1650, produced in the second half of the ninth century at Soissons or Reims, were probably written earlier, between the seventh and ninth century, in the area of Laon, Auxerre, and Reims.[55]

One of these centers may have supplied a copy of Martianus' manuscript with glosses that was brought to England in the late ninth or tenth century and that started the tradition of Martianus studies there. A commentary belonging to the group that Lutz called "Anonymus Cantabriensis" is transmitted in the late ninth- or early tenth-century manuscript Cambridge, C.C.C. 153. This manuscript contains two parts, probably put together in the second half of the tenth century. Part I has Martianus' text

---

53. On Vat. lat. 4929 see G. Billanovich, "Dall'antica Ravenna alle biblioteche umanistiche," *Annuario della Università Cattolica del Sacra Cuore (1955–57)*, (Milan, 1958), 71–107; on Heiric see E. Jeauneau, "Heiric d'Auxerre disciple de Jean Scot," in *L'école carolingienne d'Auxerre de Murethach à Remi, 830–908*, ed. D. Iogna-Prat et al. (Paris, 1991), 353–70, and other articles in this collection; on Heiric's dates see J. Contreni, "Inharmonious Harmony: Education in the Carolingian World," in *The Annals of Scholarship: Metastudies of the Humanities and Social Sciences* (New York, 1980), 1:81–96, reprinted in J. Contreni, *Carolingian Learning, Masters and Manuscripts* (Hampshire, 1992), chap. 4, esp. 87.
54. Bieler, "Text Tradition," 3–4.
55. See Szerwiniack, "Un commentaire," esp. 30–34 on the date and place of composition.

with interlinear glosses (very few on book VI), written in insular hands; part II has a running commentary to Martianus, also in an insular hand.[56] The same running commentary is transmitted in the second part of manuscript Cambridge, C.C.C. 330, written in the late ninth century in France, probably at Reims.[57] We do not know, however, when this manuscript arrived in England (the two parts were put together in the twelfth century, at the time when part I was written).[58] In addition to these two manuscripts containing the text of Anonymus Cantabriensis in the form of a running commentary, the same commentary to book VI, in the form of interlinear and marginal glosses, is transmitted in Paris, BN lat. 8669, written in the second third of the ninth century, probably at Soissons.[59] Although neither of the two French manuscripts that we have seems to have served as the immediate exemplar of the English one, (Cambridge, C.C.C. 153), they all ultimately go back to Continental manuscripts, possibly produced at Reims or Soissons.[60]

The physical appearance of the manuscripts testifies to careful and attentive study of Martianus' book VI. Glosses often seem to have consti-

---

56. T. A. M. Bishop, "The Corpus Martianus Capella," *Transactions of the Cambridge Bibliographical Society* 4 (1967): 257–75; on the hand see 262; on the codicological description see 262–64. Leonardi ("I codici," 1960, 21–22) gives the ninth century as the time of production for both part I and part II. I am grateful to Tessa Webber, who helped me analyze the handwriting of the Cambridge manuscripts.

57. I thank Rosamond McKitterick for drawing my attention to the considerable similarity between the hands of Cambridge, C.C.C. 330, pt. II, and Valenciennes, Bibl. Mun. 293. On the latter manuscript, containing Chalcidius' commentary on and translation of Plato's *Timaeus*, see R. McKitterick, "Knowledge of Plato's *Timaeus* in the Ninth Century: The Implications of Valenciennes, Bibliotheque Municipale Ms 293," in *From Athens to Chartres: Neoplatonism and Mediaeval Thought*, ed. H. J. Westra (Leiden, 1992), 85–95, reprinted in R. McKitterick, *Books, Scribes, and Learning in the Frankish Kingdoms, 6th–9th Centuries* (Aldershot, 1994), chap. 10. For the late ninth century as a possible date see Bishop, "Corpus," 267; for the eleventh century, Leonardi, "I codici," 1960, 23.

58. R. M. Thomson (in a handwritten addition in Corpus Christi College's copy of James' *Catalogue of Corpus Christi manuscripts*) identified the hand that made these glosses, as well as interlinear glosses in part II, as that of William of Malmesbury and concluded that by William's time the two parts must have been put together. See also Bishop, "Corpus," 267; Leonardi, "I codici," 1960, 22.

59. Préaux, "Les manuscrits," 79; Leonardi ("I codici," 1960, 436) cites Bernhard Bischoff's opinion on the origin. Possibly, only glosses to book VI belong to the "Anonymus Cantabriensis" group—Lutz ("Martianus," 381) placed this manuscript in the group of "unidentified glosses."

60. Bishop ("Corpus," 268) has suggested a common Continental exemplar for Cambridge, C.C.C. 153, pt. II, and C.C.C. 330, pt. II, that goes back to the ultimate archetype. Paris, BN lat. 8669, although close, was copied from a different exemplar—it contains a diagram lacking in both Cambridge manuscripts. However, my conclusions are based on the incomplete evidence of book VI.

tuted part of the manuscript design. In some manuscripts, the margins are left wide to accommodate comments.[61] In the manuscripts that I have examined, the glosses seem to have been written either at the same time as the main text or slightly later, as the gloss hands are either the same as or very similar to those of the main text.[62] Sometimes the gloss hand also makes corrections in the main text, which may have been the work of a student, corrected and annotated by a teacher.[63] Sometimes several hands write the glosses—they may have belonged to students working on Martianus' text.[64]

Thus the manuscript evidence testifies to the intensive study of Martianus Capella's book VI that began soon after the middle of the ninth century in the area of Reims and Auxerre, where scholars had also been interested in other geographical texts. From this area, the tradition of Martianus studies, including glosses to book VI, spread to other Continental centers and to England.

What was the nature of these studies? The contents of the glosses to book VI, to the analysis of which I shall now turn, can tell us about the emphases early medieval scholars made when reading Martianus' geographical text and the questions that attracted their attention.

The Contents of the Glosses

Although glosses vary in combination and contents from one manuscript to another, and though each manuscript can serve as an independent witness of teaching at a particular place and time, they all reveal certain common trends and, by implication, the common concerns that shaped these trends. Functional analysis of the glosses supports the impression made by the physical appearance of the manuscripts—that of a close attention to

---

61. Leiden, B.P.L. 88; Besançon, Bibl. Mun. 594; Paris, BN lat. 8669.

62. Glosses to book VI are written in the same hand as the main text in three manuscripts, in a similar hand in two, and in a contemporary or slightly later hand in ten.

63. Paris, BN lat. 8669.

64. In Paris, BN lat. 8671, glosses to book VI are written in several hands, one of which has made corrections in the main text; two or three hands added glosses in Leiden, B.P.L. 88. For the practice of employing students in copying manuscripts see the analysis of multiple hands in Tegernsee manuscripts in C.E. Eder, "Die Schule des Klosters Tegernsee im frühen Mittelalter im Spiegel der Tegernsee Handschriften," *Studien und Mitteilungen zur Geschichte des Benediktiner-Ordens und seiner Zweige* 83 (1972): 6–155, reprinted as a monograph, Munich, 1972. See also R. Hexter, *Ovid and Medieval Schooling: Studies in Medieval School Commentaries on Ovid's "Ars Amatoria," "Epistulae ex Ponto," and "Epistulae Heroidum"* (Munich, 1986), 151–54.

Martianus' text. In accordance with the trend displayed by other medieval glosses, all the commentators carefully explain the grammatical and lexical peculiarities of Martianus' Latin; his geographical ideas receive fewer comments. The majority of glosses supply synonyms or clarify grammatical constructions.[65] When the commentators do explain the contents, they devote much more attention to explaining the theoretical sections of book VI—the shape and size of the earth, the four elements—than to the following descriptions of places.

Shape and Size of the Earth
The commentators all agree with Martianus that the earth is neither flat nor concave but spherical.[66] In two manuscripts of the "Martin" version—Leiden, B.P.L. 88, and Leiden, Voss. lat. F. 48—the commentators hasten to make this question clear even before Martianus addresses it. Geometry hardly begins her discourse—"I am called Geometry because I have often traversed and measured out the earth, and I could offer calculations and proofs for its shape, size, position, regions, and dimensions"—when the commentators immediately add, "because it has a round shape" and "because it is situated in the lowest part of the world."[67] The commentator of Leiden, B.P.L. 88, never develops this question further. He only gives synonyms when commenting on the passage in Martianus that treats this question in detail.

By contrast, in Leiden, Voss. lat. F 48, the commentator, addressing the same passage, takes particular care to make Martianus' point as unam-

---

65. According to my rough census of Remigius' commentary (version A, edited by Lutz), which I have made following the typological method suggested by A. Schwartz and developed by R. Hexter, out of 907 glosses on the geographical part of book VI, 626 are devoted to language explanations and only 281 to contents explanations. See A. Schwartz, "Glossen als Texte," *Beiträge zur Geschichte der deutschen Sprache und Literatur* 99 (1977): 25–36; Hexter, *Ovid and Medieval Schooling*, 30–31; N. Lozovsky, "The Explanation of Geographical Material in the Commentary by Remigius of Auxerre," *Studi medievali*, 3d ser., 34 (1993): 563–72. The ratio in Remigius' commentary seems representative.

66. Martianus *De nuptiis* VI.590: *formam totius terrae non planam, ut aestimant, positioni qui eam disci diffusioris assimulant, neque concauam, ut alii, qui descendere imbrem dixere telluris in gremium, sed rotundam, globosam etiam sicut Secundus Dicaearchus asseuerat.*

67. Martianus *De nuptiis* VI.588 (trans. p. 220): *Geometria dicor, quod permeatam crebro admensamque tellurem eiusque figuram, magnitudinem, locum, partis et stadia possim cum suis rationibus explicare . . .* ; Leiden, B.P.L. 88, 292.4: *EIUSQUE FIGURAM habeat figuram quod rotundam;* ibid., 292.5: *LOCUM PARTIS quod in imo mundi est sita;* Leiden, Voss. lat. F. 48, 292.4: *FIGURAM rotunditatem.*

biguous as possible. "Nor is it concave," says Martianus. The commentator adds, "[not] like a sponge, curved, but spherical, that is, like an egg."[68] He even demonstrates this by a diagram on the margin of folio 54r, showing a circle with an inscription inside: "A disc; the earth is not like that." Below the disc is an egglike shape with an inscription inside: "The earth is like this" (fig. 1).[69]

A little later, in the margin, the commentator, anticipating Martianus' critique, warns the students not to trust Anaxagoras, who asserted that the earth was flat: "Anaxagoras, although he told the truth concerning other things, [should not be] believed in this matter."[70] Later still, when Martianus argues that if Anaxagoras were right, then on their rising celestial bodies would become immediately visible to all the inhabitants of the earth, the commentator adds a final touch: "That is, if the earth were flat."[71]

John Scottus and, following him, Remigius of Auxerre strike a different emphasis when they treat the same passage. They focus on Martianus' proof of the spherical shape of the earth. According to Martianus, "the rising and setting times of the stars would not vary according to the elevation or sloping of the earth if the operations of the heavenly firmament were spread out above a flat surface and the stars shone forth over earth and seas at one and the same time; or, again, if the rising of the sun were concealed from the hollowed cavities of the more depressed parts of the earth."[72] Paraphrasing Martianus' text in simpler words, John Scottus and Remigius clarify this argument: "If the earth were flat or concave, that is,

---

68. Leiden, Voss. lat. F. 48, 292.18: *CONCAUAM in modum spongiae in sinum sed rotundam scilicet in modum oui globosam scilicet esse.* This idea may go back to Cassiodorus, who mentioned the egg shape of the earth, naming Varro as his source: *mundi quoque figuram curiosissimus Varro sublongae rotunditati in Geometriae volumine comparavit, formam ipsius ad ovi similitudinem trahens . . .* (*Institutiones* II.7.4). I am grateful to Bruce Eastwood for the reference.

69. Leiden, Voss. lat. F. 48, fol. 54r : *Discus non ita terra; Ita terra.* The same diagram is found in Besançon, Bibl. Mun. 594, fol. 45r.

70. Martianus *De nuptiis* VI.591–92; Leiden, Voss. lat. F. 48, 293.6: *ipse Anaxagoras in aliis rebus ueritatem dixit tamen in hoc non creditur.*

71. Leiden, Voss. lat. F. 48, 293.18: *QUIN ETIAM CUNCTE scilicet si plana esset terra.* This gloss is also found in Besançon, Bibl. Mun. 594.

72. Martianus *De nuptiis* VI.591 (trans. p. 221): *namque ortus obitusque siderum non diuersus pro terrae elatione uel inclinationibus haberetur, si per plana diffusis mundanae constitutionis operibus uno eodemque tempore supra terras et aequora nituissent aut item, si emersi solis exortus concauis subductioris terrae latebris abderetur.*

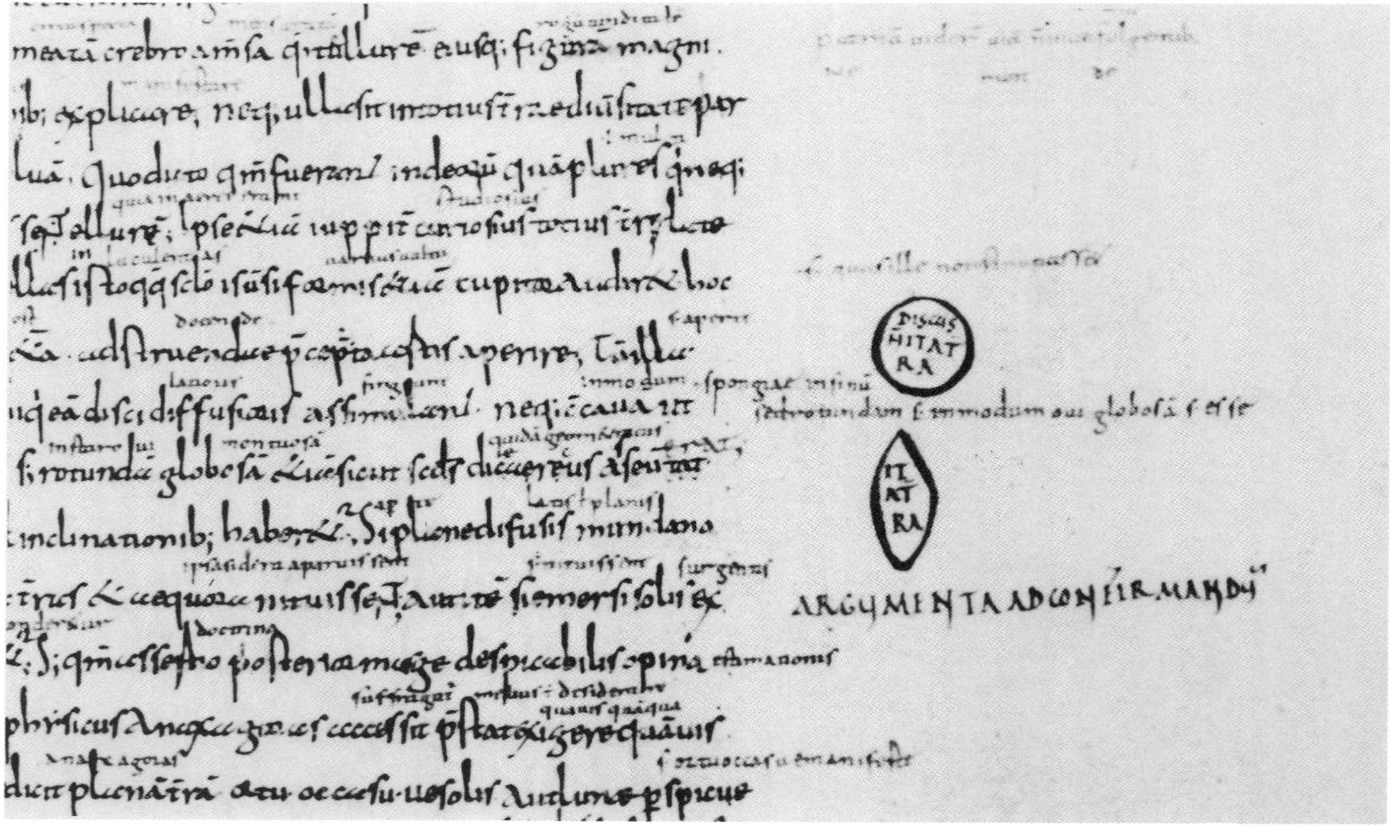

Fig. 1. Leiden, Universiteitsbibliotheek, Voss. lat. F. 48, s. IX, [Auxerre?], fol. 54r

if it were like a flat or concave disc, the rising and setting and the length of the day would be the same, which is impossible."[73]

The comments on the shape of the earth testify to the care with which the commentators treated Martianus' text rather than to any theoretical contributions. The commentators, trying to keep within the frame of Martianus' discussion, bring in no new information. The fairly straightforward passage that Martianus devoted to the shape of the earth easily allowed his commentators to explain his text using the text itself.

Martianus' passage describing how Eratosthenes had measured the earth's circumference with the help of a sundial presented more difficulties. The commentaries that I have examined, if they explain this passage at all, focus on the work of the sundial and the numbers resulting from the calculations.

The sundial and its use attract the attention of the commentator of Leiden, Voss. lat. F 48. He starts out by explaining that the sundial can be used in calculating the earth's circumference, because the shadow of the gnomon (the stick in the center of the sundial) moves around the sundial in the same way as the sun moves around the earth. In one hour, just as the sun travels through one-twenty-fourth part of the earth's circumference, the shadow in the sundial moves through one-twenty-fourth part of the clock. Next the commentator offers a method of calculating the length of this part, which is somewhat puzzling, probably due to a corrupt text. Giving the length of one part—50 stadia—he suggests that one multiply 50 by 15 to get the size of 15 parts. The rest of the gloss resembles an exercise in pure calculation, unrelated to the earth's circumference.[74]

---

73. John Scottus *Annotationes* and Remigius *Commentum*, both at 293.2: *SUBDUCTIORIS id est humilioris ac si dixisset: Si plana fuisset terra aut concava, id est si fuisset ut discus planus aut discus concavus, eodem modo omnibus esset ortus et occasus et longitudo omnium dierum, quod non potest esse.*

74. Leiden, Voss. lat. F. 48, fol. 54v, margin, near 295.19: *Quod agit umbra in horologium, hoc sol in ambitu terre. umbra autem uicessimam quartam partem horologii in una hora conlustrat. sequitur ut sol etiam uicesimam quartam partem totius terrae in una eademque hora perficiat. si autem uis nosse quae sit illa uicessima quarta pars totius ambitus terreni disci, scito quod singule partes signiferi quingenta stadia habent in arris dum infinite sunt in celesti circulo. multiplica ergo quingenta quindecies et erunt quindecim partes. adde et iam quintam partem unius partis, id est xxiv stadia, adde quartam partem unius horae et hoc totum collige, id est quingenta quindecies et quinta pars unius partis et quarta pars unius horae, et hoc totum erit uicesima quarta pars totius ambitus terreni orbis, quam uicesimam quartam partem solsticia equinoctiale horarum fecit.* The first sentence resembles John Scottus' description in *Periphyseon* III.717B: *Itaque quod umbra in horologio sensui indicat, hoc caelestium corporum incessabilem motum efficere ratio probat. . . . Sol autem idem de corporibus et terrae et stili umbram iacit. . . .* The second sentence resembles glosses by John Scottus and Remigius, with one difference—while John Scottus and Remigius talk about the water clock, the commentator of

The sundial is represented on a diagram transmitted in several manuscripts.[75] The diagram shows four concentric circles, with the gnomon as their center (fig. 2). The circles are labeled, respectively, from the largest to the smallest: "This is twenty-four"; "This is eighteen"; "This is twelve. The circle [drawn] by the top of the *stilus*"; "The circle of the shadow measures six times its length." Near the middle of the *stilus* is written, "The shadow from the middle of the *stilus.*"[76]

This diagram conveys the same information as the long gloss found in both John Scottus' and Remigius' commentaries; a manuscript containing version B of Remigius' commentary has both the gloss and the diagram.[77] The gloss explains how to calculate the circumference of circles drawn from various points of the *stilus* and, like the gloss in Leiden, Voss. lat. F 48, resembles an exercise in calculation.[78]

In Florence, San Marco 190, a long gloss (occupying the lower half of fol. 71r and the upper half of fol. 72v) follows the sundial diagram. In the middle of this gloss (on fol. 71v) is a second diagram, representing the sundial, with the gnomon in the middle of two concentric circles (fig. 3). The gloss explains how to find the circumference of the earth using the sundial.

Voss. lat. F. 48 talks about the sundial. Cf. John Scottus *Annotationes* and Remigius *Commentum*, both at 295.5: . . . *Ergo in una hora vicessima quarta pars . . . totius ambitus terrae curritur, et si multiplicaveris spatium unius partis in terra, id est D stadia, per partes signiferi, id est per CCCLX, . . . invenies ambitum totius terrae.* The remainder of the gloss by the commentator of Voss. lat. F. 48, for which I could find no parallels, is unclear to me.

75. This diagram is transmitted in manuscripts of the "Martin," "Remigius," and "Anonymus Cantabriensis" groups: Besançon, Bibl. Mun. 594, fol. 84v; Florence, San Marco 190, fol. 71r; Paris, BN lat. 8669 (the diagram on fol. 122r is unfinished); Paris, BN lat. 8671, fol. 83v; Paris, BN NAL 340, fol. 82v.

76. Florence, San Marco 190, fol. 71r: *iste vicies quater; iste decies & octies; iste duodecies. umbra ex media stili; circulus umbre est sexies illum libra metitur; circulus de summitate stili.*

77. See Paris, BN NAL 340, fol. 82v for the diagram. Vienna, Nationalbibliothek, 177 contains the same gloss, without a diagram.

78. John Scottus *Annotationes* 296.5; Remigius *Commentum* 296.4: *VICIES QUATER ac si dixisset: In aequinoctiali die umbra dimidiam partem stili tenet, ideoque dixit ESTIMATIONE CENTRI SUI. Centrum enim dicitur dimidium; igitur circulus qui ducitur a summitate umbrae, id est a medio stilo per gyrum usque ad illum locum quo coeperit, mensuratur longitudine umbrae sexies. circulus vero qui fit a summitate stili in gyro mensuratur eadem longitudine duodeties; XII ad VI duplus est. Tertium circulum mensurat decies octies, quartum circulum, id est extremum vigies quater, XX ergo IIII, id est extremus circulus ad circulum stili qui duodeties mensuratur duplus est. XX enim IIII ad XII in dupla proportione consistunt atque ideo dixit CIRCULI DUPLICIS MODUM REDDIDIT quia non dixit nisi de duobus circulis, id est de extremo et de circulo stili; ideo autem umbra sexies metitur primum circulum, quia brevissimus dies sex horas habet, ideo secundum duodeties quia aequinoctialis dies XII horas habet, ideo tertium decies octies quia longissimus dies, id est solstitialis dies, XVIII horas habet, ideo quartum vigies quater aut quia XXIIII hore sunt in ambitu totius terrae aut quia dies in Tyle XXIIII horas habet, quia nulla nox ubi est in solstitiali die.*

ab heratostene doctissimo gnomonica disputatione discussa · nuppe scaphia
dicuntur rotunda ex ere uasa · que horarū ductus stili in medio fundo sui eperi
tare discriminant · Qui stilus gnomon appellatur · Cuius umbre pluritas
equinoctio centri sui estimatione demensa · uiges quater copulata · circuli ex
pluris modu reddit · Heratostenes uero ab siene ad meroen p mensores regios
ptolomei certus de stadiorū numero redditur · quotaq portio telluris eet
aduertens · multiplicansq p partium ratione · circuli mensuraq terre incun
ctanter quot milib; stadiorū ambiretur absoluit ·

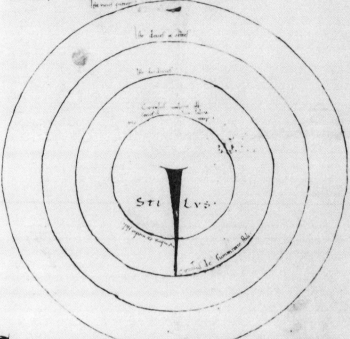

Ex horologio colligitur terre amplitudo · uicinati equinoctio · Equinoctialis enim linea que medii
terre umbram complectitur, ex equalibus spaciis supra terras subtusq nortitur · que in ··· partes di
uiditur horologica suppusitione · Ambitus enim totius spere celestis in duodecim spacia discernitur · Due
decima aut pars in ·xxx· partes resoluitur · Sed qui tota celestis spera in ·ccc·lx· horis ad eundem
punctū semp reuoluitur · necesse e ut ei duodecima pars in duab; horis equinoctialib; moueatur · Ita
gnitudo partiū celi in ipso celo ummensurabiliter quanta sit terre tenet horologica ratione
apprehenditur · quantū enim terre tenet umbrarum similitudo sine ullo discrimine tantundē et
ner pars celestis in terris · Quot enim partes sunt tot umbrarum differentie · tot horologi

Fig. 2. Florence, Biblioteca Medicea Laurenziana, San Marco 190, s.
Xex–XIin, France, fol. 71r

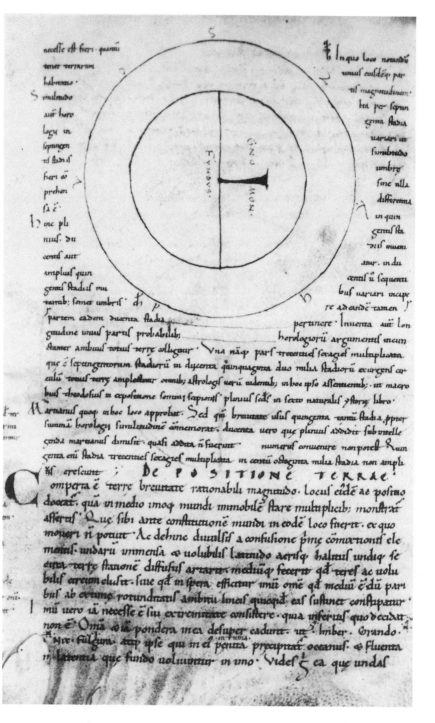

necesse est fieri quamuis
tener terrarum
habeamus·
similitudo
aut horo
logij in
septingen
tis stadiis
fieri co
prehen
sa e·
Duc pli
nius· du
centis aut
amplius quin
gentis stadiis mu
tarab; semel umbris· di
partem eadem ducenta stadia
gitudine unius partis probabilis;
tamen ambitus totius terrę colligitur· Vna nāq; pars trecenties sexagies multiplicata
que e septengentorum stadiorum in ducenta quinquaginta duo milia stadiorum excurgens circulus totius terrę amplectitur omnib; astrologis uerū uidemur; in hoc ipso assentiemur; ut macro
bius theodosius in exposicione somnii scipionis· plenius sed; in sexto naturalis ystorię libro
Marcianus quoq; in hoc loco approbat· Sed qm breuitate usus quingenta tantum stadia propter
summam horologij similitudinem commemorat· ducenta uero que plenius addidit subtrahe
cenda martianus dimisit· quasi addita non fuerint   numerus conuenire non potest Quin
genta ēm stadia trecenties sexagies multiplicata· in centum octoginta milia stadia non ampli
us crescunt ;   DE POSITIONE TERRAE·

In quo loco notandū
unius eiusdeq; par-
tis magnitudinem·
ita per septua-
genta stadia
uariari in
similitudine
umbrę
sine ulla
differentia
in quin-
genis sta
diis omnin-
tur· in du-
centis u sequenti-
bus uariari incipe
re ad euidē tamen
pertinere· Inuenta aut lon-
horologiorum argumentis meum

Comperta ē terre breuitate rationabili magnitudo· Locus eidē ae positio
docet· quia in medio imoq; mundi immobile stare multipliciб; monstrari
asseritur· Quę sibi ante constitutionē mundi in eodē loco fuerit· ex quo
moueri n potuit· Ac dehinc diuulsis a confusione primę comoctionis ele-
mentis· undarum immensa co uolubilis latitudo aerisq; halitus undiq; se
extra terrę stationē diffusus arctari· mediumq; fecerit qd teres ac uolu
bilis aeream clusit· siue qd in spera efficitur imū omē qd medium e dū partibus ab extremę rotunditatis ambitu lineis quicquid eas sustinet constipatur·
imū uero iā necesse ē siu extremitate consistere· quia inferius quo decidat
non ē· Omnia uero pondera mea desuper cadunt· ut· Imber· Grando·
Nix· fulgura· atq; ipse qui in ea penitrat precipitat oceanus· & fluenta
in latentia que fundo uoluuntur in uno· Vides ē ea quę undas

Fig. 3. Florence, Biblioteca Medicea Laurenziana, San Marco 190, s. Xex–XIin, France, fol. 71v

Although in different words, concerning the earth's circumference this commentator describes the same calculation method and gives the same numbers as John Scottus and Remigius. The length of one part on the earth multiplied by 360 gives either 252,000 stadia, according to Eratosthenes, or 180,000 stadia, according to Ptolemy, depending on the size of a stadium. The numbers given for the circumference of the earth are the same.[79]

Version B of Remigius' commentary has an explanation identical in approach but not in words to that of version A (and that of John Scottus), edited by Lutz. It is probably the most detailed and clear account of all Martianus' commentaries that I have examined. Having described the circles of the sundial, the commentator explains that the size of one part on the earth, corresponding to one part in the sky, varies from 500 to 700 stadia (however, this version, unlike version A, does not explain that the length of one stadium varies in different measurements); then he proceeds to multiply the length of one part by 360.[80]

The manuscripts of the "Anonymus Cantabriensis" group describe the same method. This commentator devotes several glosses to the earth's measurement. He begins by providing a Latin translation of the Greek words meaning "the clock."[81] Then he explains what a gnomon is: "A gnomon is a stick placed in the middle of the sundial; but according to some people this gnomon is a bigger space of 500 stadia, according to others a

---

79. Cf. John Scottus *Annotationes* 296.11 and Remigius *Commentum* 295.5. For the edition and analysis of this gloss see Leonardi, "Illustrazioni," 52–57. On the measurement of the earth in Remigius' commentary see Lozovsky, "Explanation."

80. For instance, Paris, BN NAL 340, fol. 76v, 296.5: *UICIES QUATER id est IIII circuli sunt. Primum sexies metitur umbra. Secundum duodecies. Tercium decies occies. Quartum uicies quater. Secundus ad primum duplus est. Tercius ad primum triplus est. Quartus ad primum quadruplus. Tercius ad secundum sesqualter est. Quartus ad tercium sesquitercius. Ex quarto circulo et secundo dicunt quando dicit uicies quater. Sic inuenerunt ambitum tocius terrae id est uidebant V circulos per medium coelum et uidebant ablationem umbrae ab horologio in aliquo circulo in VI[ta] hora et in crastina die ablata erat umbra in alio circulo. Ab illo horologio usque in hoc horologium D[ta] stadia in terra est, una pars in caelo secundum Plinium. Secundum Eratostenem DCC[ta] stadia in terra, una pars in celo. Septingentis igitur multiplicatis tricensies sexagies fiunt ducenta quinquaginta duo milia id est ambitus totius terrae. Quociens metitur diametrus ambitum sui circuli id est ter metitur et VII[ma] pars tercii diametri.* Cf. John Scottus *Annotationes* and Remigius *Commentum* ad loc, cited in n. 78.

81. Anonymus Cantabriensis 295.4: *HOROSCOPA id est horarum uisiones HOROSPICA id est horarum inspectiones HOROLOGICA id est horarum rationes.* Here and in subsequent quotations of this set of glosses, unless otherwise noted, I give the text according to Cambridge, C.C.C. 330, without recording the variations in spelling contained in Cambridge, C.C.C. 153 and Paris, BN lat.8669, since these variations do not change the meaning of the glosses.

smaller one of 700 stadia; and it is observed that beyond 700 stadia the sundial changes because of the earth's round shape."[82] What the commentator probably means is that the length of the gnomon's shadow changes every 500 or 700 stadia, depending on the size of a stadium. This note needs some additional explanation, which might have been supplied orally or deduced from the subsequent gloss explaining the final result of the earth's measurement: "The shadow of the sundial's *stilus* is 700 stadia of equal size. It is certain that the sun covers this distance during one day. And because the zodiac is divided into 360 parts, the sun completes its course in 360 days. Therefore, multiply 700 by 360, and you will get 252,000."[83] Like the previous gloss, this one is hard to understand without some additional comments, which may have been provided orally. Only one of the manuscripts of this group—Paris, BN lat. 8669—has a diagram representing the circles of the sundial, which, however, is not related to these particular glosses.

All the comments on the earth's measurement that I have examined share certain similarities. They always describe Eratosthenes' method as multiplication of one part of the earth by 360—very different from the geometrical procedure reported by Cleomedes, a Greek writer of the first or second century A.D.[84] The calculations and numbers always seem to attract particular attention, sometimes turning into a final goal unrelated to the earth's circumference.[85] For instance, the commentator of Leiden, B.P.L. 88, leaving Eratosthenes' method without a comment, uses Mar-

---

82. Ibid., 295.5: *GNOMINE id est gnomen est stilus in medio horologio positus. sed hic gnomen spatium est secundum primores magnos homines D stadiorum, secundum uero minores homines sicut modo sunt DCC stadiorum, et est sensus ultra DCC stadia horologium necesse est mutari propter inclinationem terrae.*

83. Ibid., 295.20: *DUOBUS MILIBUS id est umbra stili horologici per septingenta stadia uniuseiusdemque mensurae. Pro certo sit quod per singulos dies transit sol in caelo tantum. Et quia secundum zodiaci partitionem cuius singula signa trigenis partibus diuiduntur, sol cursum suum explet trecentie LX diebus. Multiplica ergo DCC per CCCLX inuenies CCLII.*

84. Cleomedes, *De motu circulari corporum caelestium*, ed. A. Todd (Leipzig, 1990), I.7. On Cleomedes' date see J. O. Thomson, *History of Ancient Geography* (New York, 1965), 215; on Eratosthenes' method as described by Cleomedes, ibid., 158–62. On Eratosthenes' method as described by Martianus Capella see W. H. Stahl, *Roman Science: Origins, Development, and Influence to the Later Middle Ages* (Madison, 1962), 174–75, and "Quadrivium," 134–35.

85. Alison White ("Glosses," 136) arrived at a similar conclusion about the glosses to Macrobius: calculations of the size of the sun, for instance, provided an opportunity for a set of exercises on arithmetic and geometry. On this practice in teaching grammar see also S. Reynolds, *Medieval Reading: Grammar, Rhetoric, and the Classical Text* (Cambridge, 1996), 75 ff.

tianus' passage about the earth's circumference (expressed in stadia) as an occasion for an exercise showing how to transfer stadia into miles: "200 stadia [equals] 25 miles. 800 stadia [equals] 100 miles. 1000 stadia [equals] 125 miles. . . ."[86] The exercise, however, seems to lack an initial explanation of how many stadia a mile contains. This information, absent in Leiden, B.P.L. 88, is provided in John Scottus' and Remigius' commentaries: "stadium is an eighth part of a Roman mile."[87] The students who read Martianus using Leiden, B.P.L. 88, probably received oral explanations from their teacher.

Martianus' description of Eratosthenes' measurement of the earth not only attracted his immediate commentators but was often excerpted and transmitted together with Macrobius' treatment of the same topic. Usually these excerpts followed one another and occurred in miscellaneous collections of astronomical materials.[88] Martianus supplied the description of the earth's measurement in the commentary to Bede's *On the Nature of Things,* transmitted in manuscript Berlin, Deutsche Staatsbibliothek Phillipps 1832 (Berlin 130), 873, Metz.[89] Martianus' passage about Eratosthenes is copied verbatim, even though Bede never mentions the earth measurement and only talks about the areas on the earth differing from each other by the length of the day and the length of the shadow the gnomon casts on the day of the equinox.[90] Regarding Martianus as an authority on the earth's measurement and the work of the sundial, this commentator tries to reconcile the difference in figures that Bede—following Pliny—and Martianus give for the length of the gnomon's shadow in Egypt. According to Pliny and Bede, this length equals slightly more than half of the gnomon's length; according to Martianus, exactly half. The

---

86. Leiden, B.P.L. 88, 295.19: *DUCENTIS Ducenta stadia XXV miliaria. Octigenta stadia centum miliaria. Mille stadia CXXV miliaria. . . .* (The text continues up to "52,000 stadia [equals] 6,500,000 miles" and ends with a total: *sint in summa triginta milia et mille miliaria quingenta* [31,100,500].)

87. John Scottus *Annotationes* 292.5 and Remigius *Commentum* 292.7: *Stadium est octava pars miliarii.*

88. For instance, in London, BL Harley 647, s. IX, France, fols. 18r–v (Martianus *De nuptiis* VIII.860 and VI.595–98); London, BL Harley 2506, s. X, Winchester, fols. 52ra–53ra (Macrobius *De somniis* I.20.14–32), fols. 53ra–va (Martianus *De nuptiis* VIII.860 and VI.595–98). For the complete list see Leonardi, "I codici," 1960.

89. The glosses from this manuscript were edited by Frances Lipp in Bede, *De natura rerum* (CCSL 123). On this manuscript see F. Lipp, "The Carolingian Commentaries on Bede's *De natura rerum*" (Ph. D. diss., Yale University, 1961), 44–99; C. Jones, introduction to Bede, *De natura rerum,* 185.

90. Bede *De natura rerum* 42: *De circulis terrae. Octo circulis terra pro dierum varietate distinguitur.* The commentator quotes Martianus *De nuptiis* VI.596–97.

commentator, worried about this difference, resolves the difficulty by saying that Pliny and Martianus describe the measurements taken in different cities—Pliny in Syene, Martianus in Meroe.[91] The commentator's high regard for and extensive use of Martianus Capella led Frances Lipp to conclude that he was connected to the circle of scholars at Laon and Auxerre, the centers of Martianus studies.[92]

Why did medieval scholars demonstrate such an interest in the earth's measurement? The commentaries to Martianus' book VI do not tell us anything about the wider context of these studies. However, John Scottus Eriugena treated this question at length in his *On Natures,* a philosophical synthesis of Christian knowledge. The context of his treatment can give us an insight into the concerns behind the calculations.

John Scottus discusses the size and measurement of the earth in book III, devoted to "the third part of universal nature" in his fourfold division of nature, the nature that is created only and does not create.[93] John Scottus organizes his treatment of the created world as a commentary on the section of the Book of Genesis that describes the first six days of creation. His discussion of the size of the heavenly bodies and the distances between them is, in fact, an extended commentary on the Fourth Day described in Genesis 1:14: "And God said, Let there be luminaries in the firmament of heaven."[94] Eratosthenes' calculations, yielding the circumference of the earth, the length of the diameter of the earth, and the distance from the earth to the moon, reveal the numerical and musical proportions of the world: "And note the prevalence in all these calculations of the perfect numbers, namely 6 and 7 and 8, which by nature constitute the chief symphonic proportion of music which is called the diapason.... For the number six multiplied by itself, that is six times six, makes 36, which, if you multiply it by 7 thousand gives you the circumference of the whole earth. For 36 times 7 thousand or 7 (thousand) times 36 makes 252,000 stades,

---

91. Phillipps 1832, 68.3, p. 231: *Martianus uero in eodem tempore meridiano aequinoctii die umbram gnominis sui diamitrum, id est medium, tenere asserit, Quid est ergo quod Plinius "paulo post quam" dicit "dimidiam gnominis mensuram efficit" umbra, Martianus uero dimidium tantum gnominis eandem umbram gnominis tenere confirmat, nisi forte non in eodem loco Aegypti comparationem gnominis et umbrae illid duo posuere sed in diuersis locis?* See the disussion in Lipp, "Carolingian Commentaries," 66.

92. Lipp, "Carolingian Commentaries," 75–76.

93. John Scottus *Periphyseon* III.23.690A, p. 189: *... ad considerationem tertiae partis uniuersae naturae de qua praedicatur creari solummodo et non creare redeundum esse censeo.* On John Scottus' fourfold division of nature see H. Bett, *Johannes Scotus Erigena: A Study in Medieval Philosophy* (New York, 1964).

94. John Scottus *Periphyseon* III.32.715A, p. 245: *Dixit autem Deus, Fiant luminaria in firmamento caeli.*

the number which comprises the girth of the whole earth."[95] Later, continuing to discuss the distances between the heavenly bodies, John Scottus cites Pythagoras as an authority on the musical proportions and adds that this idea of the structure of the world is in full accordance with the Scriptures: "he [Pythagoras] attempted to affirm by sure proofs that the structure of the whole world both rotates and is measured in accordance with musical proportions, which the divine scripture does not deny either, for it says, '[And] who will put to sleep the concert of heaven?'"[96] Thus, by studying the measurements, one approaches the secret of the world created by God, which brings one closer to understanding the Creator. According to John Scottus, even though the divine Scriptures say nothing definite "concerning such measurements of the sizes and distances of the bodies of the world," God encourages the investigation of the reasons of things visible and invisible: "For as through sense we arrive at understanding, so through the creature we return to God."[97]

The commentators of Martianus Capella do not only concentrate on numbers when discussing the size of the earth. They also emphasize calculations in other questions, unrelated to earth measurement. In some manuscripts of the "Martin" group, the commentators, addressing Martianus' passage about the division of the earth into ten climatic zones, observe: "You should note that the solstitial circle [marks] two winter regions and two more. This makes four. Similarly, the middle belt [has] four regions. Altogether there are eight. The northern and southern circles make one region each. The total is ten."[98]

---

95. Ibid., III.33.718B, p. 251: *Et uide quantum perfecti numeri in his omnibus uigent, senarius uidelicet et septenarius et octonarius, in quibus maxima simphonia musicae naturaliter constituitur, quae diapason uocatur. . . . Senarius nanque numerus per se ipsum multiplicatus, id est sex sexies, triginta sex efficit, quo numero si septem milia multiplicaueris, inuenies totius orbis ambitum. Siquidem septem milia tricies sexies aut triginta sex septies multiplicata ducenta quinquaginta duo milia stadiorum faciunt, quo numero totius telluris amplitudo includitur.*

96. Ibid., III.34.722A, p. 261: *ipse iuxta rationes musicas totius mundi fabricam et uolui et mensurari certis argumentationibus conatus est asserere, quod nec diuina negat scriptura dicens: '[Et] concentum caeli quis dormire faciet?'* (Job 18:37)

97. Ibid., III.35.723B–C, p. 263: *Et quamuis in diuinis scripturis de talibus mundanorum corporum dimensionibus magnitudinum et interuallorum nil diffinitum reperiatur, . . . diuina tamen auctoritas rationes rerum uisibilium et inuisibilium non solum non prohibet, uerum etiam hortatur inuestigari. . . . Ut enim per sensum peruenitur ad intellectum ita per creaturam reditur ad deum.*

98. In Leiden, B.P.L. 88, 298.16: *DECEM Notandum quod solstitialis arculus duas regiones brumalis item duas alias peragunt et fiunt quattuor. similiter media fascea IIII regiones cunctae simul fiunt VIII. septentrionalis circulus unam regionem sicut austrinum facit. Sunt autem omnes decem.* The same gloss, with minor variations, is found in Leiden, Voss lat. F. 48; and Besançon, Bibl. Mun. 594.

Even though the commentators do not directly refer to wider issues, their steady interest in calculations and numbers reflects and perfectly fits the mathematical context of the quadrivium, which made geometry a mathematical discipline, directly dealing with numbers. And numbers ultimately served for understanding God's creation, for, according to the Book of Wisdom, God has "ordered everything in measure, and number, and weight."[99]

Place Descriptions

Compared to the theoretical section, Martianus' descriptions of places get far less attention from his ninth- and tenth-century commentators. Only rarely do they go beyond identifying a place-name as belonging to a city, or a province, or a mountain, and so forth.[100] When the commentators do explain names, they show more interest in matters of language and literature than in geography. A place-name becomes an occasion for a grammatical exercise, where some commentators mention the declension of the name: for example, "Emus is the name of a mountain. Its genitive is Emi."[101]

More extended explanations most frequently deal with etymologies. All Martianus' commentators seem to share Isidore's conviction that etymology works as a key to understanding the thing through its name.[102] They provide translations and etymologies for Greek names, regarding this as a sufficient explanation. For instance, the only thing that all the traditions examined by me explain about Acroceraunia is the origin of its name. Several "Martin" manuscripts say: "Acroceraunia are the highest mountains of Epirus, so called from frequent lightnings because this often happens there."[103] Anonymus Cantabriensis gives a clearer etymology of the same name: "Acroceraunium, that is, 'the crest of lightnings,' because *acros*

---

99. Sapientia 11:21.

100. Here are several random examples out of a great number of these simple identifications: "Martin," Leiden, B.P.L. 88, 302.10: *PARTYENEMQUE urbs est;* cf. Remigius *Commentum* 302.10: *PARTYENEM nomen urbis;* Leiden, B.P.L. 88, 304.2: *RODUM insulam.*

101. Anonymus Cantabriensis 320.16, Paris, BN lat. 8669 fol. 80v: *EMI emus momen montis est cuius genitiuus est Emi.* Cambridge, C.C.C. 330, fol. 119r and Cambridge, C.C.C. 153, fol. 82v give a slightly different version of the same gloss: *EMUS id est mons cuius genitiuus est emi.*

102. Isidore *Etym.* I.29.2: *Nam dum videris unde ortum est nomen, citius vim ejus intelligis.*

103. Leiden, B.P.L. 88; Leiden, Voss. lat. F. 48; Florence, San Marco 190—all at 303.6: *ACROCERAUNIAM altissimi montes Epiri a crebris fulminibus dicti eo quod sepissime in illos cadant.*

means 'crest' and *ceraunos* means 'lightning.'"[104] John Scottus and Remigius (version A) combine both explanations: "*Ceraunum* means 'lightning,' *acros* 'crest,' because lightnings often fall on the crest of this mountain."[105]

Many place-names attract comments evoking mythological or literary associations. Peloponnesus is "the kingdom of Eson, the father of Jason"; Tenedum is "an island so called after Tenna, king of Troy, who had defiled his stepmother and because of this was exiled."[106] Place-names become occasions for poetic quotations, unrelated to Martianus' text but mentioning the name. For instance, Lucan's verse mentioning Lemannus (Lake Constance) travels from one commentary to another.[107]

Mythological and literary associations prevail even when it may seem appropriate to connect Martianus' text to the Christian tradition. Only once in these commentaries have I come across a biblical reference to a place-name. Contrary to the common tradition regarding the biblical river Geon flowing from Paradise as the Nile, some manuscripts of the "Martin" group identify the Ganges as the Geon. Remigius (version A) goes as far as declaring the Ganges and the Nile to be the same river, thus implicitly eliminating the contradiction: "the Ganges, that is, the Nile, which is also called the Geon."[108]

In other cases, the commentators invariably favor mythological explanations. Commenting on Martianus' passage about the division of the earth into three continents, some commentators seem to borrow their etymologies of the continents' names from Isidore. But whereas Isidore connects the three parts of the world to the biblical story about the settlement of Noah's sons, Martianus' commentators only retain mythology. Anonymus Cantabriensis treats the names of the three continents in the following

---

104. Anonymus Cantabriensis 303.6: *ACROCERAUNIUM id est summitas fulminum, ab eo quod est acros id est summitas et ab eo quod est ceraunos id est fulmen.*

105. John Scottus *Annotationes* and Remigius *Commentum,* both at 303.6: *ACROCERAUNIAM Ceraunum fulmen, acros summum, quia in summitatem illius montis cadunt saepe fulmina.*

106. Leiden, B.P.L. 88, 303.5: *PELOPONENSUM regio Esionis patris Iasonis;* Florence, San Marco 190, 303.5: *PELOPONESSUM regio Iasonis;* Remigius *Commentum* 304.3: *TENEDUM id est insula dicta a Tenne rege Troiae qui stupravit novercam suam et propterea ibi eiectus est.*

107. Leiden, Voss. lat. F. 48, and Remigius *Commentum,* both at 310.16: *deseruere cauo tentoria fixa Lemanno* (Lucan *De Bello Civili* I.396).

108. "Martin," Leiden, B.P.L. 88, and Leiden, Voss. lat. F. 48, both at 302.8: *GANGE FLUUIO Ganges ipse Geon;* Remigius *Commentum* 302.8: *GANGES id est Nilus, qui et Geon vocatur.*

way: "Three sisters, that is, Asia, Africa, and Europe, after having sex with Jupiter, obtained the whole world with its three parts for themselves, and each of them called her share by her proper name."[109] Remigius (version B) offers an explanation in the same vein, with a slight variation: "Europe and Asia were two very beautiful virgins defiled by Jupiter; and, as a present, he permitted them to call the two parts of the earth by their names, that is, Europe and Asia. But Africa is called after King Afer."[110] The reference to King Afer seems to go back to Isidore; however, Isidore connects this king to a descendant of Abraham, which Martianus' commentators do not do.

This etymological and mythological explanation of the division of the earth into the three continents seems sufficient to Anonymus Cantabriensis and the commentators of the group "Remigius-B." Many commentators—John Scottus, Remigius (version A), and the commentators of three manuscripts of the "Martin" group[111]—leave this passage without a comment. Only one manuscript provides a geographical commentary, a map of the three continents. In Florence, San Marco 190, the map on folio 74r, represents the three continents surrounded by the ocean—Asia in the top half, Europe in the lower left part, and Africa in the lower right part (fig. 4). The map also contains information about the dimensions of each part of the world and shows the boundaries dividing the continents: the river Tanais, the Riphean Mountains, and Lake Meotis between Asia and Europe; Gades between Europe and Africa; and the Nile between Asia

---

109. Anonymus Cantabriensis 306.4, Paris, BN lat. 8669, fol. 77r: *tres sorores, id est Asia, Africa et Europa, Iouis concubitui aptae totum mundum in tribus partibus sibi meruerunt et unaquaeque illarum uocauit suam partem proprio sui nomine.* Cambridge, C.C.C. 330, fol. 118v and Cambridge, C.C.C. 153, fol. 82r give a slightly different version of this gloss: *EUROPAM id est tres sorores asia europa affrica iouis concubitui aptae. Totum mundum in tribus partibus sibi meruerunt et unaqueque illarum uocauit suam partem proprio sui nomine.* Isidore's etymology of the name of Africa could have been one of the sources, but Isidore only talks about Europe abducted by Jupiter, deriving the name of Asia from the name of an eastern queen. Cf. Isidore *Etym.* XIV.4.1: *Europa quippe Agenoris regis Libyae filia fuit, quam Iovis ab Africa raptam Cretam advexit, et partem tertiam orbis ex eius nomine appellavit;* XIV.3.1: *Asia ex nomine cuiusdam mulieris est appellata, quae apud antiquos imperium tenuit orientis.*

110. Paris, BN lat. 8674, 306.4: *Europa et Asia duae uirgines fuerunt adeo pulchrae quas stupravit Iouis, et pro munere dedit eis, ut earum nominibus duae partes terrae uocarentur, id est Europa et Asia. Africa uero ab Afro rege uocata est.* On Afer cf. Isidore *Etym.* XIV.5.2: *Alii dicunt Africam appellari ab uno ex posteris Abrahae de Cethura, qui vocatus est Afer.* . . . Cf. also ibid., IX.2.115: *Afri appellati ab uno ex posteris Abrahae, qui vocabatur Afer, qui dicitur duxisse adversus Libyam exercitum, et ibi victis hostibus considisse, eiusque posteros ex nomine atavi et Afros et Africam nuncupasse.*

111. Leiden, B.P.L. 88; Leiden, Voss. lat. F. 48; and Besançon, Bibl. Mun. 594.

a gadriano eius ingressu preteriori maris p longitudine cursus uix quindecim
passuum milia numeraui. Latitudo uero ubi angustior quinq; ubi diffusa septem
ubi glacior decem milibz explicat. Hinc defluere p diuersos sinus sublidentesq;
capos tot maria tot fragores & quamui p diuersa equora tumescit. Vndarum
ii illa p ruptio interfluentis oceani leua europii facit libiaq; dextra & montib;
utriq; concluditur. Nā europa calpe africa abethna monte despicitur. Qui u
triq; eminentes dici columne herculei meruerunt. qd testimonio uetustatis labo
ris herculei limes in illis sit consecratus. siquidem ultra eu egredi consueti uel
luris inuia phibebant. Denique uia hoc de ea sacre uirtutis possibilitate p sua
sū. qd cū antea natura terris maria dispararet. ac tanto eius circuitu firma
ret oceani montiu prodictoru effossis radicib; diuulsoq; continuo camporu
deuerit lacunisq; terraru improuisum pelagus in usu impigre mortalitatis ad
misit p mutans orbis facie natureq; discrimina. Hoc igitur freto tenor su eu
ropa distendetur usq; in tanais fluminis gurgite. a quo inchoans asia. nili in
dem alueo limitat. & in quide nilus eande africaq; disrumpens telluris com
plexu intersecans multitudine fluuioru. europa tamen interminari sponi
dis faucib; dicere quia plurimi que apponitis p angusta descendens ad meo
tide quoq; perfertur.

Fig. 4. Florence, Biblioteca Medicea Laurenziana, San Marco 190, s. Xex–XIin, France, fol. 74r

and Africa. Without adding anything, this map serves as a close visual commentary of Martianus' text.[112]

Sometimes a literary association seems to be a goal in itself. When Martianus mentions Gaul, almost in passing, referring to certain stars that cannot be seen there, Anonymus Cantabriensis uses the occurrence of the name *Gaul* as an occasion for reminding the student about the divisions of Gaul going back to the Romans: "There are three Gauls: Gallia Belgica, Gallia Lugdunensis, and Gallia Aquitania."[113] This commentator leaves the actual description of Gaul later in Martianus' text without a comment. Here, not really connected to the theme of this passage, the comment resembles a literary reference, establishing the contact with other texts, rather than a geographical clarification. Would it also sound out of date for the ninth and tenth centuries? We have seen that Richer used Caesar's basic divisions but supplemented them with contemporary, more detailed information. The Martianus commentators make no such attempt.

In general, all the commentators seem to feel quite comfortable with Martianus' picture of the world, somewhat out-of-date even in his own time, not to mention the ninth or tenth century. They never try to update Martianus' text; in explaining it, they always keep within the classical image of the world. For example, Martianus repeats out-of-date information after Pliny and Solinus: "There, too, is the promontory of the Golden Horn, famous for the city of Byzantium." The name *Byzantium* is so apparently out-of-date for Martianus' own time, the fifth century, that some scholars have even tried to use this passage as evidence that Martianus did not know about Constantinople and, therefore, wrote before A.D. 330.[114] But the commentators who explain this passage at all only translate the name *Golden Horn* from Greek into Latin but never mention Constantinople.[115] They do not seem to think it appropriate to give the

---

112. See the description and discussion in Leonardi, "Illustrazioni," 48–49.

113. Martianus *De nuptiis* VI.593: . . . *cum Canopum ac Berenices crinem, stellas admodum praenitentes, Scythia Galliaeque atque ipsa prorsus non cernat Italia*. . . . Anonymus Cantabriensis 293.25: *Galliae tres sunt: gallia belgica, gallia lugdunensis et gallia aquitania*. The triple division itself may be a tribute to Julius Caesar; the names go back to Diocletian's reform. See Caesar *Bellum Gallicum* I.1 *(Gallia est omnis divisa in partes tres)* and Lugge, "'Gallia und Francia,'" 11.

114. Martianus *De nuptiis* VI.657 (trans. P. 245) : *illic promuntorium Ceras Chryseon Byzantio oppido celebratum*. . . . Cf. Pliny *Nat. hist.* IV.46 and Solinus *Collectanea* X.17. On Martianus' date see *Martianus Capella and the Seven Liberal Arts*, 2:245 n. 145, and Shanzer, *Philosophical and Literary Commentary*, 5.

115. "Martin," Florence, San Marco 190, and Anonymus Cantabriensis, all at 325.12: *CERASCHRYSEON cornu aureum;* Remigius *Commentum: CERUCRUSIUM cornu aureum; cruson enim aurum, cerus cornu.*

current name of this city here, in the geographical text. And yet, in a gloss to book V, on rhetoric, when Martianus mentions Theodore of Byzantium, a rhetorician contemporary with Plato, some commentators add, "Theodorus Byzantius, that is, from Byzantium, that is, Constantinople."[116] Would they consider this one mention sufficient for the reader to keep in mind when going from book V to book VI? This can hardly be the case. Even within book VI alone, the commentators do not rely on memory but repeat their explanations, sometimes of fairly simple words, such as *quod* (because) or *levorsum* (to the left).[117] For some reason, it seems that the commentators considered it more appropriate to update Martianus' information in the book on rhetoric than in that on geography.

Throughout Martianus' description, all the commentators seem to aim at locating places in a grammatical, mythological, or literary landscape rather than a geographical one. They may have used Martianus' information about places when they studied classical literature; Martianus' image of the world is essentially the classical one, borrowed from Pliny and Solinus.

Manuscripts of Martianus Capella's book VI accompanied by commentaries represent an influential tradition in early medieval geographical studies, demonstrating how this text was taught at Frankish and English schools. All the commentators discussed in this chapter, seeing the clarification of Martianus' text as their main goal, devoted most of their glosses to the questions of grammar, syntax, and word meaning. When they did explain the contents of book VI, the commentators distinctly favored the theoretical part of book VI over Martianus' description of places. Among other theoretical questions, numbers and calculations attracted particular attention. This concern with numbers probably reflected the goals of the quadrivium—to study the numerical proportions of the created world and thus to bring one closer to understanding the Creator.

Turning to Martianus' descriptions of places, his commentators, in the traditions of Isidore, used etymology as the main method—sometimes as the main goal—of their explanation. They never tried to update Mar-

---

116. Remigius *Commentum*, version B, 274.18: *THEODORUS BIZANTIUS id est de Bizantio, ipsa est Constantinopolis.*

117. Here are just a few examples out of many: Remigius *Commentum* 285.17, 286.6, 286.11, 286.13: *QUOD id est eo quod;* 307.8: *LEVORSUM in sinistram partem;* 330.3: *LEVORSUM in sinistra parte;* 339.8: *LEVORSUM in sinistram partem.*

tianus' picture of the world, probably perceiving it as a part of a literary or grammatical rather than geographical landscape. Their geographical space, like that of other learned geographical descriptions, is permanent, timeless, and unchanging.

Even though the commentators of book VI seem mainly to show concerns other than geographical, we should not forget that their glosses were intended to be used along with Martianus' text, which contained geographical material, not separately from it. Glosses accompanied the text and provided immediate explanation, even if only on the verbal level. The same goes for the running commentaries—even though many of them survived separately from Martianus' text, that they were sometimes bound together with this text indicates their common use.[118] The commentators seem to have trusted Martianus' geographical knowledge so much that they almost never went beyond the text of his book VI, unlike the commentator of Bede, who sometimes goes against his text, or the commentator of Orosius, who in many cases supplies information borrowed from Isidore.[119]

By closely studying Martianus' text without any attempts to change or correct his information, ninth- and tenth-century scholars accepted his classical image of the world, which went back to Pliny and Solinus. The Martianean tradition continued to study the earth of Roman geographical texts, unlike the tradition that, following Isidore, placed essentially the same classical picture into a Christian frame. Differing in their image of the world, both traditions used a similar approach to geographical studies. To compose their image of the world, they turned to texts; to explain the world, they treated it like a text.

---

118. A running commentary is bound together with Martianus' text in Cambridge, C.C.C. 330 and 153. On these manuscripts see Bishop, "Corpus."

119. On Bede's commentator see Lipp, "Carolingian Commentaries," 66–67; on Orosius' commentator, Szerwiniack, "Un commentaire."

CHAPTER 5

What Was Early Medieval Geography?

Belonging to different contexts and following different traditions, theoretical geographical knowledge in the early Middle Ages consistently displayed certain common characteristics. Drawing on various texts, biblical exegetes, historians in their introductions, encyclopedic writers, and geographers in special treatises essentially described the earth of antiquity, only rarely and occasionally including contemporary realities. Does this mean that early medieval scholars simply practiced bad geography, as many historians claim?[1] Or did medieval geographers pursue an entirely different form of knowledge, obeying rules foreign to modern geography? The present chapter, addressing these rules and characteristics, will contribute to the main question of this study: what type of knowledge was early medieval theoretical geography?

The Image of the Earth

For a modern reader, the most puzzling characteristic of early medieval theoretical geographical descriptions is their lack of correspondence to contemporary reality.[2] Orosius began his history with a long introduction representing a timeless classical picture of the world. Jordanes, Bede, and Richer evoked the same picture before turning to their respective stories.

---

1. For instance, George Kimble entitled his chapter devoted to the early Middle Ages "The Dark Ages of Geography" (*Geography in the Middle Ages,* 19). For more on the "decline thesis" see my introduction in the present study.

2. Michel Zimmerman, after analyzing the Ripoll compilation, with its formal world picture, well expressed the tantalizing question that such texts inspire ("Le monde," 72): "Comment les gens 'cultivés' pouvaient-ils s'en satisfaire? Comment accepter en effet que les termes le plus souvent employés par les contemporains pour désigner les réalités concrètes dont ils sont tributaires, soient ou bien totalement ignorés *(Francia),* ou bien utilisés dans une perspective historique...."

Isidore of Seville did not consider it appropriate to include contemporary facts in his geographical books. The ninth- and tenth-century commentators of Martianus Capella never tried to update his fifth-century text, which contained even older information.[3]

The discrepancy between the world of the classical tradition and the world of contemporary reality becomes particularly puzzling when both worlds coexist within the same work. In the eighth century, Anonymus Ravennatis, unlike many other Latin geographical writers, recorded recent political divisions. His Europe included Guasconia, the name taken by Aquitania in the seventh century.[4] He reported a more recent name for Gallia Belgica—Francia Rinensis, the name for the country of the Rhenish Franks, attested in other sources since the fifth century; he also mentioned that at his time Germania and Gallia Belgica were dominated by the Franks.[5] He recorded places not known from old Roman sources and city names in their contemporary forms.[6] At the same time, there are contemporary realities that Anonymus Ravennatis completely ignored. Describing Spain and Asian regions, he never mentioned the Muslims or the changes they had brought to the political picture of the world.

In the ninth century, Dicuil reflected contemporary realities to a certain extent. Referring to eyewitnesses whom he personally knew, he spoke of the Nile and the islands to the north of Britain.[7] At the same time, he reproduced from his sources the names and dimensions of late Roman provinces and never referred to contemporary political divisions.

How should we explain this discrepancy between the world the scholars described and the world they lived in, especially when an author combined

---

3. Orosius *Hist.* I.2; Jordanes *Getica* I.4–9; Bede *EH* I.1; Richer *Hist.* I.1. On Isidore and Martianus commentaries see above, chap. 5.

4. Anonymus Ravennatis *Cosmographia* IV.40, p. 77: *Iterum iuxta ipsam Britaniam circa limbum oceanum ponitur patria que dicitur Guasconia, que ab antiquitus Aquitania dicebatur.* On Anonymus Ravennatis' modernization see Dillemann, *La Cosmographie,* 167–80; J. Schnetz, "Zum Beschreibung des Alamannenlandes beim Geographen von Ravenna," *Zeitschrift für Geschichte des Oberrheins* 75, n.s., 36 (1921): 337–41; Staab, "Ostrogothic Geographers," 29.

5. Anonymus Ravennatis *Cosmographia* IV.24, p. 59: *. . . ponitur patria quae dicitur Francia Rinensis, que antiquitus Gallia Belgica Alobrites dicitur;* I.11, p. 10: *Prima ut hora noctis Germanorum est patria, quae modo a Francis dominatur.* On Francia Rinensis see E. Ewig, "Die Civitas Ubiorum, die Francia Rinensis und das Land Ribuarien," *Rheinische Vierteljahrsblätter* 19 (1954): 11; Lugge, *Gallia und Francia,* 160; Staab, "Ostrogothic Geographers," 40–41. On Germania and Gallia Belgica see Staab, "Ostrogothic Geographers," 41.

6. Staab, "Ostrogothic Geographers," 36, 44–45. For details on other contemporary information see ibid., with bibliography.

7. Dicuil *De mensura* VI.12–18, VII.11–13.

the two images? The discrepancy seems to stem from intention rather than poor or insufficient knowledge. Isidore chose to omit the contemporary facts about the Goths from his geographical books, reporting them elsewhere in his encyclopedia. The commentators of Martianus Capella did mention that the city of Byzantium had been given the name of Constantinople, but they chose to do so in a context other than geographical. Eighth- and ninth-century authors could not help knowing about Muslim states—Dicuil even recalled an elephant sent to Charlemagne by Harun al-Rashid—and yet he never mentioned the latter's kingdom.[8] Neither did he (or other Carolingian geographical writers) ever refer to the Carolingian Empire.

Although early medieval scholars never directly addressed this question, they must have also realized this discrepancy, apparent to them even more than to us. Writing his history in the tenth century, Richer followed Caesar's classical picture of Gaul divided into three parts. He supplemented the picture with many contemporary names. However, he did so in the course of a historical description. In his geographical introduction, he reported Caesar's divisions with no comment.

Early medieval scholars seem to have thought that geographical descriptions did not require contemporary information. By choice rather than by ignorance, they continued to reproduce the classical picture of the world, copying the writings of late antiquity, composing their own tracts, and making this image part of their education. Early medieval scholars never explained why they preferred their geographical writings to be timeless, not to say anachronistic. To understand their reasons, we need to turn to a wider context of medieval ideas about the nature of the world and knowledge about it.

How to Understand the Created World—General Ideas

For Christian scholars, who built their ideas on the foundation of Platonic philosophy, both the existence of the visible material world and knowledge about it depended on God, thus perceived as both the source and the goal of all existence and knowledge.[9] Augustine, discussing this

---

8. Dicuil *De mensura* VII.35.

9. On the often discussed connection between Platonic thought and Christianity see E. Gilson, *History of Christian Philosophy in the Middle Ages* (New York, 1955); on Augustine's Platonism see R. A. Markus, "Marius Victorinus and Augustine," in *The Cambridge History of Later Greek and Early Medieval Philosophy,* ed. A. H. Armstrong (Cambridge, 1970), 329–419, esp. pp. 364–66.

question, explained how human beings, having examined the created universe, could "arrive at the immutable being of God; and then should learn from him that everything which exists, apart from God himself, is the creation of God, and of him alone."[10] Augustine maintained that humans, because they are created in God's image, can receive this knowledge directly from God, by means of a revelation, but that most people are incapable of discovering the truth themselves. To help them, Augustine claimed, God has instituted the Scriptures, which contained all wisdom. Augustine wrote that just as in studying visible things that are out of reach of our senses we trust those who have witnessed these things, in studying invisible things we ought to "put our trust in witnesses who have learnt of these things, when they have been presented to them in that immaterial light, or who behold them continually so displayed."[11] Augustine emphasized that the text of Scripture is the most trustworthy witness for the truth of God, arguing that the creation is best testified by the words of Genesis 1:1: "In the beginning God made heaven and earth."[12] However, according to Augustine, the created world itself, the greatest of all visible things, testifies to the truth of Scripture: "The earth is our big book; in it I read as fulfilled what I read as promised in the book of God."[13]

Thus, the visible world needed to be studied. However, as Augustine, following Platonic philosophy, reminded us, bodily senses that reported information about the material world themselves belonged to the material realm and therefore should not be trusted. Augustine maintained that the raw data provided and at the same time circumscribed by bodily senses did not represent true knowledge about the world. He argued that only "the human intellect, the rational constituent of the soul of man," created by

---

10. Augustine *De civitate Dei* XI.2 (trans. p. 430): *Magnum est et admodum rarum uniuersam creaturam corpoream et incorpoream consideratam compertamque mutabilem intentione mentis excedere atque ad incommutabilem Dei substantiam peruenire et illic discere ex ipso, quod cunctam naturam, quae non est quod ipse, non fecit nisi ipse.*

11. Ibid., XI.3 (trans. p. 431): *Sicut ergo de uisibilibus, quae non uidimus, eis credimus, qui uiderunt, atque ita de ceteris, quae ad suum quemque sensum corporis pertinent: ita de his, quae animo ac mente sentiuntur . . . , hoc est de inuisibilibus quae a nostro sensu interiore remota sunt, his nos oportet credere, qui haec in illo incorporeo lumine disposita didicerunt uel manentia contuentur.*

12. Ibid., XI.4.

13. Augustine, *Epistulae*, ed. A. Goldbacher, CSEL 34–2 (Prague, Vienna, and Leipzig, 1895), 43.9.25: *maior liber noster orbis terrarum est; in eo lego completum, quod in libro dei lego promissum.*

God and placed above the material realm, could arrive at true knowledge.[14]

Augustine's approach to knowledge provided the framework for early medieval learning. In the ninth century, his ideas were further developed by John Scottus Eriugena.[15] Following Augustine, John Scottus considered God the goal and source of all knowledge. He argued that because God had revealed the truth to humans through the authority of the Scriptures, the study of the Bible represented both the main way to knowledge and the starting point in the process of reasoning.[16] In his view, the study of the visible world, God's creation and manifestation, was another way to acquire knowledge about the Creator. John Scottus explained: "if Christ at the time of His Transfiguration wore two vestures white as snow, namely the letter of the Divine Oracles and the sensible appearance of visible things, I do not clearly see why we should be encouraged diligently to touch the one in order to be worthy to find Him Whose vesture it is, and forbidden to inquire about the other, namely the visible creature, how and by what reason it is woven, I do not clearly see. For even Abraham knew God not through the letters of Scripture, which had not yet been composed, but by the revolutions of the stars."[17] Thus, according to John

---

14. Augustine *De civitate Dei* VIII.5 (trans. p. 306): *Haec mens hominis et rationalis animae natura est, quae utique corpus non est.* . . . Augustine criticizes all pagan philosophers except Plato for trusting corporeal senses and conceiving only of corporeal origins for natural phenomena. On Augustine's distrust for sensual experience see Markus, "Marius Victorinus," 374ff.

15. Of John Scottus' epistemological ideas, going back to both Latin and Greek Christian philosophical tradition, I only discuss those that he shared with Augustine and the Latin West. On John Scottus' ideas about knowledge and methods of acquiring it see A. Schneider, *Die Erkenntnislehre des Johannes Eriugena im Rahmen ihren metaphysischen und anthropologischen Voraussetzungen nach den Quellen dargestellt*, 2 vols. (Berlin and Leipzig, 1921–23); M. Cappuyns, *Jean Scot Erigène: Sa vie, son oeuvre, sa pensée* (Bruxelles, 1964), 273–315; G. Schrimpf, *Das Werk des Johannes Scottus Eriugena im Rahmen des Wissenschaftsverständnisses seiner Zeit: Eine Hinführung zu 'Periphyseon'* (Munster, 1982), 133–48. On Augustine's influence see G. Madec, "Observations sur le dossier augustinien du *Periphyseon*," in *Eriugena: Studien zu seinen Quellen, Vorträge des III. Internationales Eriugena-Colloquiums, Freiburg im Breisgau, 27.–30. August 1979*, ed. W. Beierwaltes (Heidelberg, 1980), 75–84.

16. John Scottus *Periphyseon* I.64.509A: *Sanctae siquidem Scripturae in omnibus sequenda est auctoritas quoniam in ea ueluti quibusdam suis secretis sedibus ueritas possidet*; II.15.545B: *Ratiocinationis exordium ex diuinis eloquiis assumendum esse aestimo*.

17. Ibid., III.35.723D–724A, p. 265: *Et si duo uestimenta Christi sunt tempore transformationis ipsius candida sicut nix, diuinorum uidelicet eloquiorum littera et uisibilium rerum species sensibilis, cur iubemur unum uestimentum diligenter tangere, ut eum cuius uestimentum est mereamur inuenire, alterum uero, id est creaturam uisibilem, prohibemur inquirere, et quomodo et quibus rationibus contextum sit non satis uideo. Nam et Abraham non per litteras scripturae, quae nondum confecta fuerat, uerum conuersione siderum deum cognouit.*

Scottus, in order to approach God, one should investigate and interpret not only the Scriptures but also the visible world. Investigating the latter, however, John Scottus pointed out, one should always see this as a step bringing one closer to God and, rather than study the world for its own sake, should seek eternal foundations behind the surface of visible things.[18] John Scottus reproached pre-Christian philosophers—skilled in mundane knowledge *(mundana sapientia)*—not because they made mistakes in explaining the visible world but because they did not sufficiently seek the Creator behind the creature.[19]

In the Platonic tradition of Augustine, John Scottus declared that the study of the visible world by means of the senses could give no reliable knowledge if it could not be verified by means of reason or authority. So, discussing the distances between heavenly bodies, John Scottus left the question about their validity open. He argued that natural philosophers, expressing various opinions, could never solve this matter with any certainty: because Scripture, the final source of truth, never mentioned it, the question had to remain in the realm of uncertain opinion rather than knowledge.[20] He explained that the results of observations, though useful for developing rational arguments, could never form an independent proof, "for a conjecture based on ocular observation does not serve where reason does not have a basis for argument."[21] According to this view, senses alone could never give one true knowledge about the visible world—their data had to be put in order by means of reason.[22]

---

18. Ibid., III.35.723B–C: *Non paruus itaque gradus est sed magnus et ualde utilis sensibilium rerum notitia ad intelligibilium intelligentiam. Ut enim per sensum peruenitur ad intellectum ita per creaturam reditur ad deum. Nam non sicut irrationabilia animalia solam superficiem rerum uisibilium oportet nos intueri, uerum etiam de his quae corporeo sensu percipimus rationem reddere debemus.*

19. Ibid., III.35.724A: *. . . ipsi mundanae sapientiae periti non in hoc reprehensibiles facti sunt, quasi in rationibus uisibilis creaturae errarint, sed quia auctorem ipsius creaturae non satis ultra eam quaesierint. . . .*

20. Ibid., III.33.715D: *De circulis deque interstitiis caelestium lucidissimorumque corporum multiplex uariaque sapientum mundi opinio est et ad nullam certam rationem, quantum mihi uidetur, deducta, ideoque quaesso si quid de talibus uerisimile aut ratione conueniens tibi uisum est explanare non differas;* III. 35.723A: *Hactenus de argumentationibus philosophicis mundi spatia inuestigantibus. Si cui uero haec superflua uidentur, cum sanctae scripturae testimoniis nec roborentur nec tradita sint, non nos reprehendat. Nam et ille non potest approbare haec ita non esse sicut non possumus affirmare ita esse.*

21. Ibid., III.33.721D, p. 259: *Non enim oculorum coniectura praeualet, ubi ratio sedem argumenti non habet.*

22. Ibid., V.4.868D: *Rerum namque sensibilium veram cognitionem solo corporeo sensu impossibile est inveniri.* On Eriugena's theory of cognition and on the role of senses and intellect in it see Schneider, *Die Erkenntnislehre.*

These ideas, fully developed in theological works, formed the ontological and epistemological context of early medieval knowledge about the world, including geography. Though they never address these questions in detail, early medieval geographical writers occasionally allow us rare glimpses highlighting the connections of their works to the broader context.

How to Understand and Describe the Earth—"Geographers"

Regarded as part of the knowledge about the created world, geographical learning reflected general ideas about how and why this world should be studied. For "geographers," as for theologians, God represented the source and goal of all existence and knowledge. Following Isidore of Seville and borrowing his words, Anonymus Leidensis expressed the current idea when he said that the ultimate reasons of natural phenomena, such as the tides of the ocean, "are known only to God, as the world is his creation and all the reasons of this world are known to him alone."[23] Sharing the Platonic skepticism about knowledge gained through senses, "geographers" valued an intellectual cognition of the world over an empirical one. Anonymus Ravennatis specifically emphasized that he had learned about the earth by means of intellect rather than personal experience. Without ever seeing faraway lands, such as India, Scotia, or Mauretania, he "imbibed by intellectual knowledge the [situation] of the whole world and the habitations of various nations, because this world is described in their [the philosophers'] books in the time of many [Roman] emperors."[24]

Thus, for Anonymus Ravennatis, the importance of authority, even secular, far exceeds that of experience. Compared to the highest authority of the Bible, the evidence of experience has even less weight with him: he only mentions it to demonstrate its unreliability. To prove his idea that high mountains exist beyond the ocean, he makes his imaginary opponents bring in the following objection: "Who has ever seen these mountains with his eyes? Where in the Holy Scripture do we read that they exist?" By putting the appeal to experience before the appeal to the Scripture,

---

23. Anonymus Leidensis *De situ orbis* I.3.5: *Sed hoc Deo soli cognitum est, cuius et opus mundus est, solique omnis mundi ratio nota est.* Cf. Isidore *De natura rerum* 40.

24. Anonymus Ravennatis *Cosmographia* I.1, p. 1: *nam quod apud humanum sensum possibile est: multorum phylosophorum relegi libros Christo iuvante . . . aio tibi, licet in India genitus non sim neque alitus in Scotia neque perambulaverim Mauritaniam . . . attamen intellectuali doctrina imbiui totum mundum diversarumque gentium habitationes, sicut in eorum libris sub multorum imperatorum temporibus mundus iste descriptus est.*

Anonymus emphasizes the absurdity and ignorance of the objection. Moreover, his subsequent account reveals that those who offer the objection derive it not even from their own experience but from an incorrect interpretation of the biblical text.[25]

In important theoretical questions, Anonymus Ravennatis always turns to the Bible as the source of the most reliable knowledge, to be trusted more than any other knowledge, whether gained from books or experience. For example, speaking of the location of the rivers flowing from Paradise, the Ravenna cosmographer concludes that neither the Tigris nor the Euphrates can begin in the mountains of Armenia, as some writers assert. He maintains that they flow from unknown sources, beginning in Paradise, and that to claim otherwise is to contradict the biblical text and thus endanger not only geographical truth but the very salvation of one's soul. Therefore, exhorts Anonymus Ravennatis, "making every effort to oppose whatever seems to contradict the Holy Scripture, one should daily embrace the immaculate Christian faith with an eager spirit, in order to be saved from eternal damnation and to deserve celestial joy."[26]

In the same way, Anonymus asserts the primacy of the biblical text over experience when discussing the location of Paradise. Even if someone, trying to use the evidence of senses, said that people traveling in the East might have seen Paradise, that person would be wrong, because according to the Bible, Paradise cannot be seen by human eyes and no one can travel there.[27] To discredit this (imaginary) evidence from experience further, Anonymus Ravennatis turns to the evidence of classical authority, recorded in books. Since we nowhere read that Alexander the Great ever went beyond India during his travels, even he could not have seen Paradise located far in the East.[28]

---

25. Ibid., I.10, pp. 9–10: *et quis vidit suis occulis illos montes? ubi in sancta scriptura esse leguntur? . . . Ad quos iterum obviantes illi alii phylosophi dicunt: nos nostris occulis nunquam vidimus, quia nec permittitur a deo humanis videre optutibus. nam in scriptura legimus. . . .*

26. Ibid., I.8, p. 8: *Quam ob rem nullus . . . credat . . . quod ex Armeniis montibus, quod absit, Tigris et Euphrates exurgat, sed magis magisque, quod contra sanctam scripturam apparet, totis suis cum nisibus renuat et immaculatam fidem Christianorum alacri animo omnibus diebus vite sue amplectatur, ut ab eterna dampnatione ereptus celorum gaudiis perfrui mereatur.*

27. Ibid., I.7, pp. 6–7: *. . . hoc etiam si in humana mente assederit, velut unam de iniquis cogitationibus esse decernimus, quod corruptibilis homo, qui missus fuerat ab hominibus aut sponte perambulans, suis corporalibus occulis potuisset vel modo possit nobilissimum videre paradisum aut pollutis terram sanctam perambulare pedibus.*

28. Ibid., I.8, p. 7: *sed et in eodem libro Alexandri . . . nullo modo invenitur, quod Alexander mare transisset et sic usque in eremum, quae est trans totam Indiam, perambulasset. . . .*

Anonymus Leidensis, mentioning both experience and intellectual cognition as sources of knowledge, emphasizes that the latter is more reliable. Even though the ocean can be studied "by both corporeal eyes and the internal mental glance," he argues, "he who perceives in his mind seas and lands not seen by every glance and teaches others will be much renowned."[29] According to Anonymus Leidensis, although we can learn about the seas both from "the experience of many people" and from "the probable documents [composed by] Martianus Felix Capella,"[30] the latter inspire more trust. Turning to the description of the seas, Anonymus immediately quotes Martianus Capella as the trustworthy source that recorded the previous experience. In the prologue, Anonymus Leidensis declares only authority to be the main source of his work, composed "so that it might briefly expose the face of the earth according to the vigorous explanation of [various] authors."[31] For Anonymus Leidensis, the information brought by experience only seems trustworthy if it has been recorded and thus validated by written tradition.[32]

The same wish to validate the information gained from experience seems to have motivated Jordanes when he carefully placed his account about Scandza, possibly going back to oral sources, in the chain of classical tradition. Jordanes apparently succeeded in making his description of Scandza, for which he could only find scarce information in classical sources, part of geographical tradition—Anonymus Ravennatis in the eighth century and Freculf of Lisieux in the ninth quoted Jordanes as the "wisest cosmographer" and a respectable authority on Scandza.[33]

In light of the general attitude of early medieval scholars toward authority and experience, the case of Dicuil seems very unusual. Unlike

---

29. Anonymus Leidensis *De situ orbis* I, *Praefatio* 1: *Oceanum extima telluris mirabiliter cingentem suisque fluoribus medias spatiatim terras trifaria divisione permeantem, qui tam oculis carneis quam interno mentis intuitu nosse desiderat, qualiter multorum id relationibus declaratum sit, quantotius explorare nullatenus abnuat, quatenus tam maria quam terras non per omnia lumine visas animo cernat aliosque docens multorum laudetur affatu.*

30. Ibid. I.5.1: *Hunc igitur Oceanum multis nominibus pro regionum diversitate taxatum circumquaque navigabilem cum plurimorum didicerimus experimentis, tum Marciani Felicis Capellae probabilibus documentis....*

31. Ibid., *Proemium* 3: *...faciem terrarum auctorum vigente ratione succincte aliquantula demonstratione patefaceret.*

32. In the sphere of early medieval law, written evidence seems to have been also valued more than oral: see R. McKitterick, *The Carolingians and the Written Word* (Cambridge, 1989), esp. 60–75.

33. Jordanes *Getica* III.16–24; Anonymus Ravennatis *Cosmographia* I.12, p. 11; Freculph of Lisieux *Chronicon* I.2.16.961B–C. Also see above, chap. 3.

other geographical writers, who relied solely on written authority, Dicuil included several eyewitness reports. Speaking of the Nile, he referred to a monk named Fidelis who had told Dicuil's teacher Suibneus, in Dicuil's presence, that a branch of the Nile flows into the Red Sea.[34] Dicuil also reported what Fidelis and his companions saw when sailing along the Nile on their way to Jerusalem, including pyramids that they had called "the seven barns built by holy Joseph, according to the number of the years of abundance."[35] Writing about the islands to the north of Britain, Dicuil recalled the account of certain clerics about the length of the day in Thule and its nonfreezing sea.[36] He also described northern islands that were reported to him by a priest but that no written authority has ever mentioned.[37] Finally, he corrected Julius Solinus, recalling an elephant sent to Charlemagne by Harun al-Rashid: "Iulius [Solinus] . . . makes one mistake about elephants when he says that the elephant never lies down, for he certainly does lie down like an ox, as the people at large of the Frankish kingdom saw the elephant at the time of emperor Charles."[38]

Introducing these accounts, Dicuil seems to doubt whether his audience would consider their information valid enough against written authority—that of Julius Solinus.[39] Just as Jordanes had done in the case of Scandza, Dicuil carefully placed his eyewitness reports in the chain of written tradition. In the passage about the Nile, he first cited Pliny, then Solinus, after which he introduced Fidelis' information: "Although we read in no author's book that a branch of the Nile flows into the Red Sea, yet brother Fidelis asserted this and related it, in my presence, to my

---

34. Dicuil *De mensura* VI.12, p. 63.

35. Ibid., VI.12–18, p. 63: *deinde in Nilo longe nauigando septem horrea secundum numerum annorum habundantiae, quae sanctus Ioseph fecerat.* . . . This identification of the pyramids as Joseph's barns has become traditional in early medieval literature (cf. Gregory of Tours *Historiae* I.10; see also Tierney's note in Dicuil, *Dicuili liber*, 112 n. 13, and W. Bergmann, "Dicuils *De mensura orbis terrae*," esp. 531).

36. Dicuil *De mensura* VII.11–13.

37. Ibid., VII.14–15.

38. Ibid., VII.35, p. 83: *Sed idem Iulius nuntiando de Germania insulisque eius unum de elephantibus mentiens falso loquitur dicens elephantem numquam iacere, dum ille sicut bos certissime iacet, ut populi communiter regni Francorum elephantem in tempore imperatoris Karoli uiderunt.*

39. On Dicuil's critical approach to authority see Bergmann, "Dicuils"; on the eyewitness accounts introduced as part of the critique of Solinus see Lozovsky, "Carolingian Geographical Tradition," 39–41.

Anonymus Leidensis, mentioning both experience and intellectual cognition as sources of knowledge, emphasizes that the latter is more reliable. Even though the ocean can be studied "by both corporeal eyes and the internal mental glance," he argues, "he who perceives in his mind seas and lands not seen by every glance and teaches others will be much renowned."[29] According to Anonymus Leidensis, although we can learn about the seas both from "the experience of many people" and from "the probable documents [composed by] Martianus Felix Capella,"[30] the latter inspire more trust. Turning to the description of the seas, Anonymus immediately quotes Martianus Capella as the trustworthy source that recorded the previous experience. In the prologue, Anonymus Leidensis declares only authority to be the main source of his work, composed "so that it might briefly expose the face of the earth according to the vigorous explanation of [various] authors."[31] For Anonymus Leidensis, the information brought by experience only seems trustworthy if it has been recorded and thus validated by written tradition.[32]

The same wish to validate the information gained from experience seems to have motivated Jordanes when he carefully placed his account about Scandza, possibly going back to oral sources, in the chain of classical tradition. Jordanes apparently succeeded in making his description of Scandza, for which he could only find scarce information in classical sources, part of geographical tradition—Anonymus Ravennatis in the eighth century and Freculf of Lisieux in the ninth quoted Jordanes as the "wisest cosmographer" and a respectable authority on Scandza.[33]

In light of the general attitude of early medieval scholars toward authority and experience, the case of Dicuil seems very unusual. Unlike

---

29. Anonymus Leidensis *De situ orbis* I, *Praefatio* 1: *Oceanum extima telluris mirabiliter cingentem suisque fluoribus medias spatiatim terras trifaria divisione permeantem, qui tam oculis carneis quam interno mentis intuitu nosse desiderat, qualiter multorum id relationibus declaratum sit, quantotius explorare nullatenus abnuat, quatenus tam maria quam terras non per omnia lumine visas animo cernat aliosque docens multorum laudetur affatu.*

30. Ibid. I.5.1: *Hunc igitur Oceanum multis nominibus pro regionum diversitate taxatum circumquaque navigabilem cum plurimorum didicerimus experimentis, tum Marciani Felicis Capellae probabilibus documentis. . . .*

31. Ibid., *Proemium* 3: *. . . faciem terrarum auctorum vigente ratione succincte aliquantula demonstratione patefaceret.*

32. In the sphere of early medieval law, written evidence seems to have been also valued more than oral: see R. McKitterick, *The Carolingians and the Written Word* (Cambridge, 1989), esp. 60–75.

33. Jordanes *Getica* III.16–24; Anonymus Ravennatis *Cosmographia* I.12, p. 11; Freculf of Lisieux *Chronicon* I.2.16.961B–C. Also see above, chap. 3.

other geographical writers, who relied solely on written authority, Dicuil included several eyewitness reports. Speaking of the Nile, he referred to a monk named Fidelis who had told Dicuil's teacher Suibneus, in Dicuil's presence, that a branch of the Nile flows into the Red Sea.[34] Dicuil also reported what Fidelis and his companions saw when sailing along the Nile on their way to Jerusalem, including pyramids that they had called "the seven barns built by holy Joseph, according to the number of the years of abundance."[35] Writing about the islands to the north of Britain, Dicuil recalled the account of certain clerics about the length of the day in Thule and its nonfreezing sea.[36] He also described northern islands that were reported to him by a priest but that no written authority has ever mentioned.[37] Finally, he corrected Julius Solinus, recalling an elephant sent to Charlemagne by Harun al-Rashid: "Iulius [Solinus] . . . makes one mistake about elephants when he says that the elephant never lies down, for he certainly does lie down like an ox, as the people at large of the Frankish kingdom saw the elephant at the time of emperor Charles."[38]

Introducing these accounts, Dicuil seems to doubt whether his audience would consider their information valid enough against written authority—that of Julius Solinus.[39] Just as Jordanes had done in the case of Scandza, Dicuil carefully placed his eyewitness reports in the chain of written tradition. In the passage about the Nile, he first cited Pliny, then Solinus, after which he introduced Fidelis' information: "Although we read in no author's book that a branch of the Nile flows into the Red Sea, yet brother Fidelis asserted this and related it, in my presence, to my

---

34. Dicuil *De mensura* VI.12, p. 63.
35. Ibid., VI.12–18, p. 63: *deinde in Nilo longe nauigando septem horrea secundum numerum annorum habundantiae, quae sanctus Ioseph fecerat.* . . . This identification of the pyramids as Joseph's barns has become traditional in early medieval literature (cf. Gregory of Tours *Historiae* I.10; see also Tierney's note in Dicuil, *Dicuili liber,* 112 n. 13, and W. Bergmann, "Dicuils *De mensura orbis terrae,*" esp. 531).
36. Dicuil *De mensura* VII.11–13.
37. Ibid., VII.14–15.
38. Ibid., VII.35, p. 83: *Sed idem Iulius nuntiando de Germania insulisque eius unum de elephantibus mentiens falso loquitur dicens elephantem numquam iacere, dum ille sicut bos certissime iacet, ut populi communiter regni Francorum elephantem in tempore imperatoris Karoli uiderunt.*
39. On Dicuil's critical approach to authority see Bergmann, "Dicuils"; on the eyewitness accounts introduced as part of the critique of Solinus see Lozovsky, "Carolingian Geographical Tradition," 39–41.

teacher Suibneus. . . . ."[40] By this introduction, Dicuil both asserted that he thought his oral source was valid, equal to a written book, and recognized that his audience may have thought otherwise. When speaking of the northern islands, Dicuil also placed the oral account within the written tradition. He reported the information borrowed from Pliny, Isidore, Priscian, and Solinus before the clerics' account. By citing his oral sources along with the written ones, Dicuil both demonstrated his regard for the classical tradition and asserted the validity of his eyewitness report, reliable enough to support his critique of the written authority: "Therefore [that is, on the basis of the eyewitness report] those authors are wrong and give wrong information, who have written that the sea will be solid about Thule. . . ."[41]

Still, ultimately Dicuil may have valued classical authority more than eyewitness accounts. He hastened to corroborate Fidelis' information with that found in a classical author: "Today, in the *Cosmography* which was made in the consulate of Julius Caesar and Marcus Antonius, I have found a branch of the Nile described as flowing into the Red Sea. . . ."[42] Concluding his description of northern islands, he seems to apologize for not having found a proper confirmation: "I have never found these islands mentioned in any book by any author."[43]

Dicuil's use of eyewitness information is in fact rather limited. Like other early medieval geographical writings, his treatise essentially relies on written authority. And like his contemporaries, he makes sure to connect his work to the previous tradition, pointing out that he wrote "according to the authority of the men whom the holy emperor Theodosius had sent to measure the said provinces," supplementing this information "on the high authority of Plinius Secundus."[44]

Other "geographers" demonstrated their concern with tradition in the

---

40. Dicuil *De mensura* VI.12, p. 63 (trans. modified by me): *Quanquam in libris alicuius auctoris fluminis Nili partem in Rubrum mare exire nequaquam legimus, tamen affirmans Fidelis frater meo magistro Suibneo narrauit coram me.* . . .

41. Ibid., VII.13, p. 75: *Et idcirco mentientes falluntur qui circum eam [Thule] concretum fore mare scripserunt.* . . .

42. Ibid., VI.20, p. 65: *Hodie in Cosmographia quae sub Iulio Caesare et Marco Antonio consulibus facta est scriptam inveni partem Nili fluminis exeuntem in Rubrum mare.* . . .

43. Ibid., VII.15, p. 77 (trans. modified by me): *Numquam eas insulas in libris auctorum memoratas inuenimus.*

44. Ibid., *Prologus* 1, p. 45: . . . .*secundum illorum auctoritatem quos sanctus Theodosius imperator ad prouintias praedictas mensurandas miserat; et iuxta Plinii Secundi praeclaram auctoritatem.* . . .

same way, listing their sources in the very beginning. The anonymous author of the *Situs orbis terre* refers to works by Christian authors—"the book of the blessed presbyter Orosius and the book of the lord bishop Esodore *[sic]*."[45] Their authority guarantees for the reader that he will find the subsequent account trustworthy: "if you wish to know without error the location of the whole earth and [its] various regions, . . . you will find it in the history below."[46] Anonymus Leidensis offers his reader the same guarantee, naming his sources—Pomponius Mela, Aethicus the cosmographer, Martianus Capella, Julius Solinus, Orosius, and Isidore.[47]

Respect for authority, among other characteristics of early medieval geographical knowledge, became a subject for parody in the *Cosmography* of Aethicus Ister. In this work, the anonymous author claims to be Jerome, translating the work of the philosopher Aethicus. Addressing the subject of his book, "Jerome" declares that it is largely based on Aethicus' own experience and that, moreover, Aethicus had addressed many questions omitted by Moses in the Old Testament. To clear himself of possible accusations of temerity, "Jerome" points out that he only repeats bold theories of other people.[48] Thus the anonymous author establishes in the beginning of his work the pattern of his parody: Aethicus, the philosopher, proposes bold and unusual theories about the world while "Jerome," the translator, criticizes them.[49] By pointing out that Aethicus' knowledge lies outside biblical scholarship, the author mockingly reverses the established scholarly pattern of his time and indicates that Aethicus' knowledge

---

45. *Situs orbis terre, Praefatio: de libro beati Orosi presbyteri siue de libro domni Esodori episcopi.*

46. Ibid.: *Si scire uis absque errore totius orbis diuersarumque regionum situs, . . . per subter adnotatam istoriam repperies qualitatem.*

47. Anonymus Leidensis *De situ orbis, Proemium* 2: *Melam Pomponium dico atque Aethicum cosmografum, Martianum Felicem Capellam, Solinum Polistoriarum, Orosium necnon Isidorum ceteraque quam plurima argumenta.*

48. Aethicus Ister *Cosmographia*, pp. 87–88: *Illi [philosophi] conati sunt tam magna dixisse, quae nos metuendo ac dubitando scribere vel legere in usum coepimus temeranter adtrectare, quur Aethicus iste chosmografus tam difficilia appetisse didiceret, quaeque et Moyses et vetus historia in enarrando distulit et hic secerpens protulit. Unde legentibus obsecro, ne me temerario aestiment, cum tanta ob aliorum audacia mea indagatione cucurrisse quae conpererint.*

49. On the *Cosmography* as parody see Hillkowitz, *Zur Kosmographie*, 1:73, and the analysis in Löwe, "Ein literarische Widersacher." Even though scholars have largely rejected the main arguments proposed by Heinz Löwe (the identification of the author with Virgil of Salzburg, who had supposedly written the *Cosmography* as part of his polemic on the question of the antipodes), most of them seem to agree about the parodical nature of the *Cosmography*. On "Jerome's" critique of Aethicus see Löwe, "Salzburg als Zentrum," 125–26. On the two personae see Herren, "Wozu diente die Fälschung."

should not be taken too seriously. By criticizing Aethicus' "bold theories"—which have no support in the Bible, the ultimate source of all knowledge—the author emphasizes his point.

Not only did Aethicus go beyond biblical knowledge, but he also reported many things entirely unknown to any authority, classical or recent, pagan or Christian. Speaking of the northern island Rifargica, unknown from any other source, the author emphasizes that Aethicus "wrote many other things about this island that our predecessors *[maiores nostri]* either did not know or did not want to disclose."[50] In the same way, the crude customs of the people living on the Caspian shore (they eat bloodstained flesh of prematurely born animals and birds, believe in divination, and worship the sun and the moon) had been unknown to anyone before Aethicus reported them: "both the writers of sacred books and our predecessors in our codices omit this *[a maioribus omissa sunt]*. . . ."[51] In mentioning "predecessors," the author of *Cosmography* seems to allude to Orosius, a well-known authority who had begun his geographical exposition with reference to his predecessors, classical authors *(maiores nostri)*.[52] These allusions, mocking a scholarly treatise, seem to indicate that Aethicus' "unknown" facts belong to the realm of fiction. Every time the author of *Cosmography* mentions that the information Aethicus reported is otherwise unknown, he seems to be warning readers not to take the information seriously. Sometimes the author makes this clear by directly stating that he doubts whether what Aethicus reported is true, as he does in speaking of the people descending from Iaphet and living in the north: "he reported much about the unknown people in this place that seems dubious. . . ."[53]

The author's emphasis on unknown facts gained from experience rather

---

50. Aethicus Ister *Cosmographia*, p. 133: *Multa et alia philosophus de hac insola scribit, quae maiores nostri aut ignoraverunt aut noluerunt patefacere.* On the name *Rifargica*, probably invented by analogy with the Riphean Mountains, see Hillkowitz, *Zur Kosmographie*, 2:143 n. 228; Prinz's introduction to Aethicus Ister, *Die Kosmographie*, 21.

51. Aethicus Ister *Cosmographia*, pp. 168–69: *Carnes animalium et bestiarum et cuncta abortiva et morticina cruenta in usum vescentur. Auguria vel avium voces in diis colentes adorant solem ac lunam. . . . Haec dementia gentium illarum inaudita et incognita a nobis esse debetur vel ab scriptoribus sacrorum librorum et in codicibus nostris ideo a maioribus omissa sunt.* . . .

52. Orosius *Hist.* I.2.1: *Maiores nostri orbem totius terrae, oceani limbo circumsaeptum, triquadrum statuere.* . . .

53. Aethicus Ister *Cosmographia*, p. 154: *Nunc vero de ignotis gentibus huic loco multa praedixit, quod credere dubium est, de Iafeth et quae in plaga septentrionale conmorare vel cohabitare scribens.* . . .

than authority reverses the fundamental rules of early medieval theoretical geographical knowledge. Just as the very possibility of parodying a geographical tract demonstrates a certain status of geographical knowledge, the reversal of its rules not only reveals but also asserts their existence.

With authority rather than experience providing their only reliable and trustworthy knowledge about the world, early medieval geographical writers were mainly concerned with constructing a world image that corresponded to the biblical and classical picture rather than contemporary reality. Modifications and developments in this theoretical knowledge were achieved within the same system of reference. "Geographers" changed their image as suited their purposes by operating with authority. Different uses of the same source could produce works very different in direction. Both the Ripoll compiler and the anonymous author of the *Situs orbis terre* used Isidore as their main source. However, the former concentrated almost exclusively on Isidore's etymologies, whereas the latter, ignoring etymologies, only retained Isidore's geographical descriptions.[54] Dicuil, in his *On the Measurement of the Earth,* looking for measurements everywhere, even used Isidore's *Etymologies* as a source for measurements and numbers.[55] Dicuil's *On the Measurement of the Earth* and another Carolingian tract, Anonymus Leidensis' *On the Location of the Earth,* demonstrate a very sophisticated approach to their sources. They change and update their picture by carefully compiling their texts, omitting or slightly changing the information of their sources.[56]

However, even with updates and changes, theoretical geography essentially conveyed the timeless image of the earth, Roman in its descriptions of regions. In constructing this image, early medieval scholars turned to authority that they believed provided the most reliable knowledge about the world. Geographical tradition belonged to the areas of early medieval learning that valued written, classical authority over experience.

The Uses of Theoretical Geographical Knowledge

Formal geographical knowledge in the early Middle Ages followed its own principles, very different from those that people used in their everyday life.

---

54. For the Ripoll compiler see Zimmerman, "Le monde"; for the *Situs orbis terre* see Gautier Dalché, *"Situs orbis terre,"* 160–61; see also above, chap. 4.

55. Lozovsky, "Carolingian Geographical Tradition," 38.

56. On their changes in the world image see Gautier Dalché, "Tradition"; on their critical approach to sources, Bergmann, "Dicuils."

Information drawn from classical books could hardly help in practical situations, such as travels or foundations of monasteries. For instance, when Sturmi, the founder of Fulda, was searching for a place to start a monastery in the 740s, his own travels, not books, provided this "avid explorer of places" with the information he needed. Reporting to Boniface, his bishop, Sturmi described to him "the location of the place, and the quality of the land, and the course of water, and the [water] sources, and the valleys, and everything that this place had."[57] In his explorations, Sturmi used his own practical experience as well as the practical experience of other people—his *Life* mentions that once during his travels, he met a man who told him place-names and showed him streams.[58] Practical considerations based on experience clearly dominated Sturmi's search for a suitable place, his final decision, and its approval by Boniface.

The considerable practical knowledge that early medieval people undoubtedly possessed about places belonged to a sphere quite different from the literary geographical tradition that we see in their written works. This separation of the two spheres, which seems intentional, highlights the fact that formal geographical studies in the early Middle Ages represented a theoretical branch of knowledge that had its own rules and uses.

Useless in everyday practice, theoretical geographical knowledge served entirely different purposes, mainly defined by the ultimate goal of medieval studies of the natural world—to understand God by means of understanding his creation. Geographical knowledge regarded the earth as a physical testimony of God's work. As such, the earth could become a subject of contemplation, meditation on the questions of earthly transience and human sins—an image and purpose demonstrated by biblical commentaries. The earth could become the background and participant of human events, inviting the contemplation of human fate and divine providence. Orosius, the first historian to use geography in this way, emphasized the contemplative significance of his classical geographical image, preparing the reader for the subsequent description of "the conflicts of the human race and the world. . . ."[59] The earth could become a text that

---

57. Eigil, *Die Vita Sturmi des Eigil von Fulda,* ed. P. Engelbert (Marburg, 1968), 7, p. 138: *avidus locorum explorator;* ibid., 5, p. 135: . . . *et loci positionem et terrae qualitatem et aquae decursum et fontes et valles et omnia quae ad locum pertinebant, per ordinem exposuit.*

58. Ibid., 8 p. 140: . . . . *coepit ei locorum nomina indicare et torrentum et fontium fluenta denuntiare.*

59. Orosius *Hist.* I.1.14 (trans. p. 5): . . . *in quantum ad cognitionem uocare suffecero, conflictationes generis humani et ueluti per diuersas partes ardentem malis mundum face cupiditatis incensum e specula ostentaturus necessarium reor.* . . .

required explanation and interpretation; Isidore of Seville and Hrabanus Maurus accomplished this by means of etymological and allegorical approach. The earth could be simply described, seemingly for its own sake and with no other goals in mind, as in many geographical tracts. Sometimes, however, an author gives us an indication that his description may have pursued other goals.

This may be the case with Anonymus Leidensis, who in the beginning of his work pointed out that the Northmen's raids on the Gallic coast had served as an incentive for his writing: "Since in the past and now, the shores of Gaul close to the Mediterranean Sea and also its inner regions, because many rivers permeate them, were devastated by the horrible cruelty of the Northmen, I, admonished by the wish of certain of our brothers who do not know much about these sea raids and inspired by the gentle requests of my students, have tried to compose this little book on the location of the earth. . . ."[60] However, in describing Gaul, Anonymus Leidensis draws on Julius Caesar and Pomponius Mela, faithfully reproducing the tripartite division of Gaul from the first century B.C.[61] How would this classical image help Anonymus' brothers who wanted more information because of the recent invasions? Anonymus apparently did not mean to give any precise contemporary locations; his intentionally classical picture may have been meant to help people contemplate the disastrous consequences of the Northmen's raids.

Thus providing an image to contemplate and interpret, not a travel guide to follow step by step, theoretical geographical knowledge apparently satisfied the needs of early medieval scholars. Within their system of learning, geography had its own uses that both defined and explained its main characteristic—the lack of interest in contemporary realities.

Theoretical geographical knowledge in the early Middle Ages, largely ignoring contemporary reality, essentially reproduced the image of the world familiar since antiquity. This seemingly puzzling characteristic

---

60. Anonymus Leidensis *De situ orbis, Proemium* 1: *Cum olim quidem ut nunc Galliarum litora mari oceano mediterraneoque inminentia necnon earundem aliquatenus media, videlicet quo variis fluviorum permeantur alveis, dira Normannorum vastarentur saevitia, studio quorundam fratrum nostrorum admonitus, immo ob utriusque maris aliquantulum ignotos navigationis excursus, discipulorum mitissima deprecatione accensus, hunc de situ orbis libellum . . . componere studui. . . .*

61. Ibid., II.3.9–14, esp. II.3.14: *Gallia est omnis divisa in partes tres, quarum unam incolunt Belgae, aliam Aquitani, tertiam qui ipsorum lingua Celtae, nostra Galli appellantur* (= Caesar *Bellum Gallicum* I.1).

resulted from the fact that geographical writers shared medieval ideas about the created world and its cognition: the earth had to be studied not for its own sake but for a better understanding of the Creator. Regarding the earth as a testimony of the divine work rather than as a final goal of their studies, early medieval scholars described and interpreted this testimony using methods that they believed could provide true knowledge. Distrusting senses and sensual experience, they relied on the work of intellect represented in authoritative texts that belonged to either the Christian or the classical tradition. The resulting geographical knowledge had a purely theoretical character, serving purposes other than the practical needs of everyday life. Providing material for contemplation and education, it responded to the needs of early medieval literary and biblical learning.

Conclusion

The picture of theoretical geographical knowledge that emerges from the chapters of this book demonstrates that this knowledge, not fitting into the modern definition of geography, represented a unique phenomenon of medieval culture. In theory it lacked a definite disciplinary status and boundaries, but in practice it was perceived as an important and, to a certain extent, self-sufficient subject. Without being directly represented in early medieval classifications of knowledge, it was regarded as belonging to the field of knowledge about the created world. According to educational programs of that time, this knowledge aimed at preparing people for biblical studies, so that they might achieve the ultimate goal of Christian learning—to gain knowledge about God. Pursuing this ultimate goal and developing within this program, geographical knowledge established its own methods and rules.

Of the many contexts in which geographical knowledge functioned, exegesis was the most important, providing both its goal and its inspiration. The reciprocal relations between biblical studies and geography worked on many levels, influencing the contents and methods of both fields. In constructing their image of the world, generations of early medieval scholars transferred exegetical paradigms to geographical writings and used them again in exegesis. The modes of perception and description of the world demonstrated in exegesis informed and shaped geographical knowledge.

History, a subject tightly connected to geography, provided another context for its development. In the four histories that this study has addressed, theoretical geographical tradition, represented in introductions, helped the historians establish the background context of human events, thus supporting their various goals. Throughout their histories, the descriptions of places, also serving these goals, received different forms. Histories devoted to recent or contemporary events but using the classical geographical framework filled it in with contemporary information. Thus the classical image of the world either fully represented or only indicated in the introduction could differ from the picture emerging from the subsequent historical account.

The analysis of geographical studies at school, represented by two influential traditions, demonstrates how geographical knowledge acquired and developed some of its essential characteristics—abstract character, lack of interest in contemporary reality, application of methods of textual analysis to the natural world, the background presence of nongeographical concerns. Transmitted and promoted by means of education, these characteristics shaped theoretical geography so consistently and to such an extent that they may be considered rules defining a certain type of knowledge.

An image of the world corresponding to the classical tradition rather than contemporary reality appears to be a goal of early medieval geographical studies as much as a by-product of their authors' methods. The ontological and epistemological ideas of the time, shared by authors of theoretical geographical texts, defined their methods of collecting and representing their information. Knowledge about the earth, based on written tradition rather than immediate experience, had a deliberately speculative and abstract character.

Attempts to represent this speculative tradition as geography, with all the modern implications of the word, would lead us to the inevitable conclusion that this discipline declined in the early Middle Ages; looking for geography in the Middle Ages could be, indeed, "a disheartening study."[1] Only by thinking of this tradition as an integral part of medieval culture, tightly connected to other areas of learning and corresponding to the needs of medieval society, can we begin to understand early medieval geographical knowledge. Restored to its proper context, this knowledge, in its turn, becomes a window on medieval culture and society. Thus the speculative, theoretical, learned geographical tradition of the Middle Ages, analyzed in this study, can provide insights into medieval people's teaching practices, their attitudes toward written and oral authority, their epistemological and ontological ideas. The theoretical tradition represented in texts, however, was only a facet of a larger phenomenon: geographical

---

1. This is the conclusion John Kirtland Wright reached about medieval geography in *Geographical Lore*, 361.

knowledge of that time, which also included, for example, mapmaking and practical geographical skills. This study only touched on some of the important issues involved in the search for medieval geographical knowledge; much more remains to be done.

The early medieval geographical tradition needs to be studied in the broader context of medieval learning and literature; this would involve parallels to other fields, such as grammatical studies, which demonstrate the same reciprocal relation to exegesis as geography; or historiography, which had the same uncertain disciplinary status; or ethnography, which often used geographical descriptions when explaining the origins or movements of peoples.[2] Parallels also need to be extended into other periods and cultures.[3] A comprehensive study of geographical knowledge in antiquity and the later Middle Ages would reveal the broader outlines and perspectives of its development; a comparative analysis of medieval European and, for instance, Chinese and Japanese geographical knowledge would contribute to our understanding of the mentality behind verbal and cartographic representations of space used by early cultures.[4]

These future studies will prepare the common ground for the conceptual framework and terminology necessary for accurately describing phenomena like the one that this study deals with. The fluidity and uncertainty of terms I have tentatively used here ("theoretical geographical knowledge," "geographical tradition," "geographical studies") reflect both the lack of contemporary scholarly consensus and the more fundamental problem of method. Once we reject the modern definition of geography,

---

2. I thank John Contreni, who brought the point about grammar to my attention in a personal communication, and one of my anonymous readers, who raised the questions about historiography and ethnography.

3. A recent study of the relations between science and Daoism in China demonstrates similar problems that the students of early sciences face and proposes a new conceptual framework that can be applied to the Western context: see A. Volkov, "Science and Daoism: An Introduction," *Taiwanese Journal for Philosophy and History of Science* 8 (1996–97): 1–58.

4. Interesting and suggestive parallels in the ways medieval Europeans and early modern Japanese represented both real and imagined space in maps and texts were revealed at a panel entitled "Maps, Words, and Other Worlds: Reassessing Cartographic Traditions in Medieval Europe and Early Modern Japan" at the meeting of the American Historical Association, Pacific Coast Branch, in August 1998. I am grateful to the organizer, participants, and discussants of this panel, namely, to Marcia Yonemoto, Robert Eskildsen, Victoria Morse, Christine Petto, and Gail Bernstein, for a stimulating and fruitful discussion.

Important advances in creating the conceptual framework of the study of ancient Chinese geography, made by Vera Dorofeeva-Lichtmann, can also be tested in the European context; see V. V. Dorofeeva-Lichtman, "Conception of Terrestrial Organization in the *Shan hai jing*," *Bulletin de l'École Française de l'Extrême-Orient* 82 (1995): 57–110.

we face the question of how to focus and delineate our search. The solution that I propose—to take as a starting point the things medieval scholars understood as and included in their *orbis descriptio,* largely as represented in special "geographical" treatises—is not ideal. First, it is implicitly based on the modern idea of a special treatise in a well-defined field. Second, it limits the subjects of analysis: for instance, if early medieval "geographers," unlike classical ones, did not normally include ethnographical descriptions, does this mean that ethnographical questions did not belong to the body of geographical knowledge? Third, this choice of focus considerably narrows the number of medieval sources that deal with representations of space, maps being the most glaring exclusion. Future studies will surely propose some alternative approaches; meanwhile, it is my hope that this book's inquiry into the nature and contexts of early medieval geographical tradition will be a step toward creating a conceptual framework that would allow us to understand and describe other premodern "sciences," with their specific ways of studying the natural world.

descriptions of places, also serving these goals, received different forms. Histories devoted to recent or contemporary events but using the classical geographical framework filled it in with contemporary information. Thus the classical image of the world either fully represented or only indicated in the introduction could differ from the picture emerging from the subsequent historical account.

The analysis of geographical studies at school, represented by two influential traditions, demonstrates how geographical knowledge acquired and developed some of its essential characteristics—abstract character, lack of interest in contemporary reality, application of methods of textual analysis to the natural world, the background presence of nongeographical concerns. Transmitted and promoted by means of education, these characteristics shaped theoretical geography so consistently and to such an extent that they may be considered rules defining a certain type of knowledge.

An image of the world corresponding to the classical tradition rather than contemporary reality appears to be a goal of early medieval geographical studies as much as a by-product of their authors' methods. The ontological and epistemological ideas of the time, shared by authors of theoretical geographical texts, defined their methods of collecting and representing their information. Knowledge about the earth, based on written tradition rather than immediate experience, had a deliberately speculative and abstract character.

Attempts to represent this speculative tradition as geography, with all the modern implications of the word, would lead us to the inevitable conclusion that this discipline declined in the early Middle Ages; looking for geography in the Middle Ages could be, indeed, "a disheartening study."[1] Only by thinking of this tradition as an integral part of medieval culture, tightly connected to other areas of learning and corresponding to the needs of medieval society, can we begin to understand early medieval geographical knowledge. Restored to its proper context, this knowledge, in its turn, becomes a window on medieval culture and society. Thus the speculative, theoretical, learned geographical tradition of the Middle Ages, analyzed in this study, can provide insights into medieval people's teaching practices, their attitudes toward written and oral authority, their epistemological and ontological ideas. The theoretical tradition represented in texts, however, was only a facet of a larger phenomenon: geographical

---

1. This is the conclusion John Kirtland Wright reached about medieval geography in *Geographical Lore,* 361.

knowledge of that time, which also included, for example, mapmaking and practical geographical skills. This study only touched on some of the important issues involved in the search for medieval geographical knowledge; much more remains to be done.

The early medieval geographical tradition needs to be studied in the broader context of medieval learning and literature; this would involve parallels to other fields, such as grammatical studies, which demonstrate the same reciprocal relation to exegesis as geography; or historiography, which had the same uncertain disciplinary status; or ethnography, which often used geographical descriptions when explaining the origins or movements of peoples.[2] Parallels also need to be extended into other periods and cultures.[3] A comprehensive study of geographical knowledge in antiquity and the later Middle Ages would reveal the broader outlines and perspectives of its development; a comparative analysis of medieval European and, for instance, Chinese and Japanese geographical knowledge would contribute to our understanding of the mentality behind verbal and cartographic representations of space used by early cultures.[4]

These future studies will prepare the common ground for the conceptual framework and terminology necessary for accurately describing phenomena like the one that this study deals with. The fluidity and uncertainty of terms I have tentatively used here ("theoretical geographical knowledge," "geographical tradition," "geographical studies") reflect both the lack of contemporary scholarly consensus and the more fundamental problem of method. Once we reject the modern definition of geography,

---

2. I thank John Contreni, who brought the point about grammar to my attention in a personal communication, and one of my anonymous readers, who raised the questions about historiography and ethnography.

3. A recent study of the relations between science and Daoism in China demonstrates similar problems that the students of early sciences face and proposes a new conceptual framework that can be applied to the Western context: see A. Volkov, "Science and Daoism: An Introduction," *Taiwanese Journal for Philosophy and History of Science* 8 (1996–97): 1–58.

4. Interesting and suggestive parallels in the ways medieval Europeans and early modern Japanese represented both real and imagined space in maps and texts were revealed at a panel entitled "Maps, Words, and Other Worlds: Reassessing Cartographic Traditions in Medieval Europe and Early Modern Japan" at the meeting of the American Historical Association, Pacific Coast Branch, in August 1998. I am grateful to the organizer, participants, and discussants of this panel, namely, to Marcia Yonemoto, Robert Eskildsen, Victoria Morse, Christine Petto, and Gail Bernstein, for a stimulating and fruitful discussion.

Important advances in creating the conceptual framework of the study of ancient Chinese geography, made by Vera Dorofeeva-Lichtmann, can also be tested in the European context; see V. V. Dorofeeva-Lichtman, "Conception of Terrestrial Organization in the *Shan hai jing*," *Bulletin de l'École Française de l'Extrême-Orient* 82 (1995): 57–110.

Selected Bibliography

Manuscript Sources

I list here only the manuscripts that I have consulted, either in libraries or on microfilm.

Besançon, Bibliotheque Municipale, 594, s. IX 2/2, [St. Oyan?].
Cambridge, Corpus Christi College, 153, pt. I, s. IXex–Xin, Wales; pt. II, s. X 3/4, England.
Cambridge, Corpus Christi College, 330, pt. I, s. XI–XII, England; pt. II, s. IXex, [Reims?].
Cologne, Dombibliothek, 193, s. X–XI.
Florence, Biblioteca Medicea Laurenziana, San Marco 190, s. Xex–XIin, France.
Leiden, Universiteitsbibliotheek, B.P.L. 36, s. IX, [Lorsch?].
Leiden, Universiteitsbibliotheek, B.P.L. 87, s. IX.
Leiden, Universiteitsbibliotheek, B.P.L. 88, s. IX 2/2, [Reims?].
Leiden, Universiteitsbibliotheek, Voss. lat. F. 48, s. IX, [Auxerre?].
Leiden, Universiteitsbibliotheek, Voss. lat. F. 113 I–II, s. IX.
Paris, Bibliotheque Nationale de France, lat. 4892, s. XII.
Paris, Bibliotheque Nationale de France, lat. 8669, s. IX, [Soissons?].
Paris, Bibliotheque Nationale de France, lat. 8670, s. IX, Corbie.
Paris, Bibliotheque Nationale de France, lat. 8671, s. IX 2/3, [Auxerre?].
Paris, Bibliotheque Nationale de France, lat. 8674, s. X, Auxerre.
Paris, Bibliotheque Nationale de France, NAL 340, s. X, Cluny.
St. Petersburg, Publichnaya Biblioteka, Clas. lat. F.V. 10, s. X, Corbie.
Vatican, Bibliotheca Vaticana, Reg. lat. 1535, s. IX, [Auxerre?].
Vatican, Bibliotheca Vaticana, Reg. lat. 1987, s. IX, [Reims?].
Vatican, Bibliotheca Vaticana, Vat. lat. 4929, s. IX.
Vienna, Nationalbibliothek, 177, s. X.

Edited Sources

Adamnan. *De locis sanctis.* Ed. Ludwig Bieler. CCSL 175, pp. 175–234. Turnhout, 1965.

Aethicus Ister. *Die Kosmographie des Aethicus.* Ed. Otto Prinz. Munich, 1993.
Alcuin. *Interrogationes et responsiones in Genesin.* PL 100, 515–66.
Ambrose. *De Abraham.* Ed. C. Schenkl. CSEL 32–1, pp. 498–638. Vienna, 1897.
———. *Exameron,* ed. C. Schenkl. CSEL 32–1, pp. 1–261. Vienna, 1897.
Ammianus Marcellinus. *Rerum gestarum libri qui supersunt.* Vol. 1. Ed. V. Garthausen. 1874. Reprint, Stuttgart, 1968.
Angelomus of Luxeuil. *Commentarius in Genesin.* PL 115, pp. 107–244.
The Anglo-Saxon Missionaries in Germany. Trans. C. H. Talbot. London, 1981.
Anonymus Leidensis. *De situ orbis libri duo.* Ed. Riccardo Quadri. Padua, 1974.
Anonymus Ravennatis. *Cosmographia.* In *Itineraria Romana,* ed. Joseph Schnetz, 1: 1–110. Stuttgart, 1990.
Augustine. *Concerning the City of God against the Pagans.* Trans. H. Bettenson. New York, 1972.
———. *De civitate Dei.* Ed. B. Dombart and A. Kalb. 4th ed. Teubner, 1878–79. Reprint, CCSL 47–48, Turnhout, 1955.
———. *De doctrina Christiana.* Ed. J. Martin. CCSL 32, pp. 1–167. Turnhout, 1962.
———. *De Genesi ad litteram.* Ed. J. Zycha. CSEL 28–1. Vienna, 1894.
———. *De Genesi contra Manichaeos.* PL 34, 173–220.
———. *De Trinitate.* Ed. W. J. Mountain and F. Glorie. CCSL 50A. Turnhout, 1968.
———. *Epistulae.* Ed. A. Goldbacher. CSEL 34–2. Prague, Vienna, and Leipzig, 1895.
———. *The Literal Meaning of Genesis.* Trans. John H. Taylor. 2 vols. New York, 1982.
———. *On Christian Doctrine.* Trans. D. W. Robertson. New York, 1958.
———. *Quaestionum in Heptateuchum libri VII.* Ed. I. Fraipont. CCSL 33, pp. 1–377. Turnhout, 1958.
Bede. *Bede's Ecclesiastical History.* Ed. Bertram Colgrave and R. A. B. Mynors. Oxford, 1969.
———. *De locis sanctis.* Ed. I. Fraipont. CCSL 175, pp. 247–80. Turnhout, 1965.
———. *De natura rerum.* Ed. C. W. Jones, CCSL 123, pp. 174–234. Turnhout, 1975.
———. *Libri quatuor in principium Genesis.* Ed. Charles W. Jones. CCSL 118A. Turnhout, 1967.
———. *Nomina locorum ex beati Hieronimi presbiteri et Flavi Iosephi collecta opusculis.* Ed. David Hurst. CCSL 119, pp. 273–87. Turnhout, 1962.
———. *Nomina regionum atque locorum de Actibus Apostolorum.* Ed. M. L. W. Laistner. CCSL 121, pp. 167–78. Turnhout, 1983.
———. *Venerabilis Baedae opera historica.* Ed. Charles Plummer. 2 vols. in 1. 1896. Reprint, Oxford, 1966.
Bernard. *Itinerarium Bernardi, monachi Franci.* In *Descriptiones terrae sanctae,* ed. T. Tobler, 85–99. Leipzig, 1874. Reprint, Hildesheim, 1974.
Boethius. *De institutione arithmetica.* Ed. Gottfried Friedlein. Leipzig, 1867.
Caesar, Julius. *Commentarii rerum gestarum.* Vol. 1, *Bellum Gallicum.* Ed. W. Hering. Leipzig, 1987.

Cassiodorus. *Cassiodori Senatoris Institutiones.* Ed. R. A. B. Mynors. Oxford, 1937.

———. *An Introduction to Divine and Human Readings by Cassiodorus Senator.* Trans. Leslie W. Jones. New York, 1969.

Christian Druthmar. *Expositio in Matthaeum evangelistam.* PL 106, 1261–1504.

Cicero. *Letters to Atticus.* Ed. and trans. David R. Shackleton Bailey. Cambridge, 1965.

Cleomedes. *De motu circulari corporum caelestium.* Ed. A. Todd. Leipzig, 1990.

Dicuil. *Dicuili liber de mensura orbis terrae.* Ed. J. J. Tierney and Ludwig Bieler. Dublin, 1967.

Dunchad. *Glossae in Martianum.* Ed. Cora Lutz. Lancaster, 1944.

The Earliest Life of Gregory the Great, by an Anonymous Monk of Whitby. Ed. and trans. Bertram Colgrave. Lawrence, Kans., 1968.

Egeria. *Itinerarium Egeriae.* Ed. A. Francescini and R. Weber. CCSL 175, pp. 29–90. Turnhout, 1965.

Eigil. *Die Vita Sturmi des Eigil von Fulda.* Ed. Pius Engelbert. Marburg, 1968.

Eutropius. *Breviarium ab urbe condita.* Ed. H. Droysen. MGH AA II. Berlin, 1961.

Flodoard. *Annales.* Ed. P. Lauer. Paris, 1905.

The Fragmentary Classicising Historians of the Later Roman Empire. Ed. and trans. R. C. Blockley. 2 vols. Liverpool, 1981–83.

Freculph of Lisieux. *Chronicon.* PL 106, 917–1258.

Geographi Latini minores. Ed. Alexander Riese. Heilbronn, 1878.

Geometria incerti auctoris. In *Gerberti postea Silvestri II papae Opera Mathematica (972–1003),* ed. Nicolai Bubnov, 310–65. Hildesheim, 1963.

Gesta sanctorum patrum Fontenellensis coenobii. Ed. F. Lohier and R. P. J. Laporte. Rouen and Paris, 1936.

Gildas. *De excidio et conquestu Britanniae.* Ed. Theodor Mommsen. MGH AA XIII, pp. 1–85. Berlin, 1898. Reprint, Berlin, 1961.

Gregory of Tours. *Historiae.* Ed. Bruno Krusch. MGH SRM I-1. Hanover, 1951.

Gregory the Great. *Dialogues.* Ed. Adalbert de Vogüé. 2 vols. Paris, 1979.

Haimo of Auxerre. *Commentarius in Genesin.* PL 131, 51–134.

———. *Homiliae de tempore.* PL 118, 11–746.

Horace. *Opera.* Ed. David R. Shackleton Bailey. Stuttgart, 1991.

Hrabanus Maurus. *Commentariorum in Genesim libri quattuor.* PL 107, 439–670.

———. *Commentariorum in Matthaeum libri octo.* PL 107, 727–1156.

———. *Rabani Mauri de universo libri viginti duo.* PL 111, 9–614.

Isidore. *Differentiae.* PL 83, 9–98.

———. *Historia Gothorum.* Ed. Theodor Mommsen. MGH AA XI, pp. 241–95. Berlin, 1894. Reprint, 1961.

———. *Isidore de Séville, Traité de la nature.* Ed. Jacques Fontaine. Bordeaux, 1960.

———. *Isidore of Seville's History of the Goths, Vandals, and Suevi.* Trans. Guido Donini and Gordon B. Ford. Leiden, 1970.

———. *Isidori Hispalensis episcopi Etymologiarum sive Originum libri XX.* Ed. W. M. Lindsay. 2 vols. Oxford, 1911.

Jerome. *De situ et nominibus locorum Hebraicorum.* In *Onomastica sacra,* ed. Paul de Lagarde, 118–90. Göttingen, 1887.
———. *Hebraicae quaestiones in libro Geneseos.* Ed. Paul de Lagarde. CCSL 72, pp. 1–56. Turnhout, 1959.
———. *In Hiezechielem.* Ed. F. Glorie. CCSL 75. Turnhout, 1964.
John Scottus. *De divisione naturae.* PL 122, 439–1022.
———. *Iohannis Scotti Annotationes in Marcianum.* Ed. Cora Lutz. Cambridge, 1939.
———. *Iohannis Scotti Eriugenae Periphyseon (De diuisione naturae).* Vols. 1–3, ed. and trans. I. P. Sheldon-Williams and Ludwig Bieler, and vol. 4, ed. Edouard A. Jeauneau with the assistance of Mark A. Zier, trans. John J. O'Meara and I. P. Sheldon-Williams. Dublin, 1968–95.
Jordanes. *The Gothic History of Jordanes.* Trans. Charles Christopher Mierow. 2d ed. Princeton, 1915. Reprint, Cambridge and New York, 1966.
———. Iordan, *O proishozhdenii i deyaniyah Getov: Getica* (On the Origin and Deeds of the Getae: Getica). Ed. and trans. Elena Ch. Skrzhinskaya. Moscow, 1960.
———. *Romana et Getica.* Ed. Theodor Mommsen. MGH AA V, pp. 1–138. Berlin, 1882.
Martianus Capella. *De nuptiis Philologiae et Mercurii.* Ed. A. Dick. Corrected by Jean Préaux. 1925. Reprint, Stuttgart, 1988.
———. *De nuptiis Philologiae et Mercurii.* Ed. J. Willis. Leipzig, 1983.
———. *The Marriage of Philology and Mercury.* Trans. William H. Stahl, Richard Johnson, and E. L. Burge. In *Martianus Capella and the Seven Liberal Arts,* vol. 2. New York, 1977.
Mela, Pomponius. *De situ orbis.* Ed. C. Frick. Leipzig, 1880.
Orosius. *Historiarum aduersum paganos libri VII.* Ed. Karl Zangemeister. CSEL 5. Vienna, 1882. Reprint, Hildesheim, 1967.
———. *Le storie contro i pagani.* Ed. A. Lippold. 2 vols. Milan, 1976.
———. *The Seven Books of History against the Pagans.* Trans. Roy J. Deferrari. Washington, 1964.
Paul the Deacon. *Historia Langobardorum.* Ed. Ludwig Bethmann and Georg Waitz. MGH SRL, pp. 7–219. Hanover, 1878.
Pliny. *Historia naturalis.* Ed. Karl Mayhoff. 6 vols. 1898. Reprint, Stuttgart, 1967–70.
Pseudo-Aethicus. *Cosmographia.* In *Geographi Latini minores,* ed. Alexander Riese, 71–103. Heilbronn, 1878.
Remigius of Auxerre. *Remigii Autissiodorensis Commentum in Martianum Capellam.* Ed. Cora Lutz. Leiden, 1962.
Richer. *Histoire de France.* Ed. and trans. Robert Latouche. 2 vols. Paris, 1967.
———. *Richeri historiarum libri IIII.* Ed. Georg Waitz. 2d ed. MGH SRG 51. Hanover, 1877.
Solinus. *Collectanea rerum memorabilium.* Ed. Theodor Mommsen. Berlin, 1895.
Versus de Asia et de universi mundi rota. Ed. F. Glorie. CCSL 175, pp. 433–54. Turnhout, 1965.

Modern Authors

Amsler, Mark. *Etymology and Grammatical Discourse in Late Antiquity and the Early Middle Ages.* Amsterdam and Philadelphia, 1989.

Anspach, A. E. "Das Fortleben Isidors im VII bis IX Jahrhundert." In *Miscellanea Isidoriana, Homenaje a S. Isidoro de Sevilla en el XIII centenario de su muerte,* 323–56. Rome, 1936.

Backes, H. *Die Hochzeit Merkurs und der Philologie: Studien zu Notkers Martian-Übersetzung.* Sigmaringen, 1982.

Barré, Henri. *Les homéliaires carolingiens de l'école d'Auxerre.* Vatican, 1962.

Bately, J. M., and D. A. Ross. "A Check List of Manuscripts of Orosius' *Historiarum adversus paganos libri septem.*" *Scriptorium* 15 (1961): 329–34.

———. "The Relationship between Geographical Information in the Old English Orosius and Latin Texts other than Orosius." *Anglo-Saxon England* 1 (1972): 45–62.

Beazley, Charles R. *The Dawn of Modern Geography: A History of Exploration and Geographical Science.* Vol. 1, *From the Conversion of the Roman Empire to A.D. 900, with an Account of the Achievements and Writings of the Christian, Arab, and Chinese Travellers and Students.* London, 1897.

Becker, Gustav. *Catalogi bibliothecarum antiqui.* Bonn, 1885.

Bergmann, Werner. "Dicuils *De mensura orbis terrae.*" In *Science in Western and Eastern Civilization in Carolingian Times,* ed. Paul. L. Butzer and Dietrich Lohrmann, 527–37. Basel, Boston, and Berlin, 1993.

Berschin, Walter. *Greek Letters and the Latin Middle Ages: From Jerome to Nicolas of Cusa.* Trans. Jerold C. Frakes. Washington, 1988.

Bett, Henry. *Johannes Scotus Erigena: A Study in Medieval Philosophy.* New York, 1964.

Bevan, W. L., and H. W. Philott. *Mediaeval Geography: An Essay in Illustration of the Hereford Mappa Mundi.* London, 1873.

Bieler, Ludwig. "The Text Tradition of Dicuil's *Liber de mensura orbis terrae.*" *Proceedings of the Royal Irish Academy* 64 C 1 (1965): 1–31.

Billanovich, Giuseppe. "Dall'antica Ravenna alle biblioteche umanistiche." *Annuario della Università Cattolica del Sacra Cuore (1955–57)* (Milan, 1958), 71–107.

Bischoff, Bernhard. "Die europaeische Verbreitung der Werke Isidors von Sevilla." In *Isidoriana,* 317–44. Leon, 1961. Reprinted in Bernhard Bischoff, *Mittelalterliche Studien: Ausgewählte Aufsätze zur Schriftkunde und Literaturgeschichte,* vol. 1 (Stuttgart, 1966), 171–94.

———. *Mittelalterliche Studien: Ausgewählte Aufsätze zur Schriftkunde und Literaturgeschichte.* 3 vols. Stuttgart, 1966–81.

Bischoff, Bernhard, and Michael Lapidge. *Biblical Commentaries from the Canterbury School of Theodore and Hadrian.* Cambridge, 1994.

Bishop, T. A. M. "The Corpus Martianus Capella." *Transactions of the Cambridge Bibliographical Society* 4 (1967): 257–75.

Borst, Arno. *Das Buch der Naturgeschichte: Plinius und seine Leser im Zeitalter des Pergaments.* Heidelberg, 1994.

Bowlus, Charles R. *Franks, Moravians, and Magyars: The Struggle for the Middle Danube, 788–907.* Philadelphia, 1995.

Brincken, Anna-Dorothee von den. "Mappa mundi und Chronographia." *Deutsches Archiv für Erforschung des Mittelalters* 24 (1968): 118–86.

———. "'. . . ut describeretur universus orbis': Zur Universalkartographie des Mittelalters." *Miscellanea Mediaevalia* 7 (1970): 249–78.

Brodersen, Kai. *Terra Cognita: Studien zur römischen Raumerfassung.* Hildesheim, Zurich, and New York, 1995.

Brown, Peter. *The Cult of the Saints: Its Rise and Function in Latin Chrisitianity.* Chicago, 1981.

Brunhölzl, Franz. *Geschichte der lateinischen Literatur des Mittelalters.* Vol. 1. *Von Cassiodor bis zum Ausklang der karolingischen Erneuerung.* Munich, 1975.

Burrows, Toby. "Holy Information: A New Look at Raban Maur's *De naturis rerum.*" *Parergon* 5 (1987): 28–27.

The Cambridge History of the Bible. Ed. S. L. Greenslade. 3 vols. Cambridge, 1969–70.

Cappuyns, Maïeul. "Cassiodore." In *Dictionnaire d'histoire et de géographie ecclésiastique,* vol. 11, cols. 1349–1408. Paris, 1949.

———. *Jean Scot Erigène: Sa vie, son oeuvre, sa pensée.* Bruxelles, 1964.

Cardman, Francine. "Fourth-Century Jerusalem: Religious Geography and Christian Tradition." In *Schools of Thought in the Christian Tradition,* ed. Patrick Henry, 49–64. Philadelphia, 1984.

Cassiodoro: Dalla corte di Ravenna al Vivarium di Squileace: Atti del Convegno Internazionale di Studi, Squileace, 25–27 Ottobre 1990. Ed. Sandro Leanza. Soveria Mannelli, 1993.

Cohen, Jeremy. *"Be Fertile and Increase, Fill the Earth and Master It": The Ancient and Medieval Career of a Biblical Text.* Ithaca and London, 1992.

Cohn, Robert. *The Shape of Sacred Space: Four Biblical Studies.* Chico, Calif., 1981.

Colish, Marcia L. "Carolingian Debates over *Nihil* and *Tenebrae*: A Study in Theological Method." *Speculum* 59 (1984): 757–95.

Contreni, John. *Carolingian Learning, Masters and Manuscripts.* Hampshire, 1992.

———. *The Cathedral School of Laon from 850 to 930: Its Manuscripts and Masters.* Munich, 1978.

———. "Inharmonious Harmony: Education in the Carolingian World." In *The Annals of Scholarship: Metastudies of the Humanities and Social Sciences,* 1:81–96. New York, 1980. Reprinted in John Contreni, *Carolingian Learning, Masters and Manuscripts* (Hampshire, 1992), chap. 4.

———. "Three Carolingian Texts Attributed to Laon." *Studi medievali,* 3d ser., 17 (1976): 802–13.

Corsini, Eugenio. *Introduzione alle "Storie" di Orosio.* Turin, 1968.

Courcelle, Pierre. "Histoire d'un brouillon cassiodorien." *Revue des études anciennes* 44 (1942): 65–86.

———. *Late Latin Writers and Their Greek Sources.* Trans. Harry E. Wedeck. Cambridge, Mass., 1969.

———. "La vision cosmique de saint Benoît." *Revue des études augustiniennes* 13 (1967): 97–117.

Cristiani, Marta. "Le problème du lieu et du temps dans le livre Ier du 'Periphyseon.'" In *The Mind of Eriugena: Papers of a Colloquium, Dublin, 14–18 July 1970,* ed. John J. O'Meara and Ludwig Bieler, 41–47. Dublin, 1973.

The Cultural Context of Medieval Learning: Proceedings of the First International Colloquium on Philosophy, Science, and Theology in the Middle Ages, September 1973, ed. John E. Murdoch and Edith D. Sylla. Boston, 1975.

Curtius, Ernst Robert. *European Literature and the Latin Middle Ages.* Trans. Willard R. Trask. 1953. Reprint, New York, 1963.

Dagenais, John. *The Ethics of Reading in Manuscript Culture: Glossing the "Libro de buen amor."* Princeton, 1994.

Dagron, Gilbert. "Discours utopique et récit des origines." Pt. 1, "Une lecture de Cassiodore-Jordanès: Les Goths de Scandza à Ravenne." *Annales: Économies, sociétés, civilizations* 26 (1971): 290–305.

Daniélou, Jean. "Terre et paradis chez les Pères de l'Eglise." *Eranos Jahrbuch* 22 (1953–54): 432–72.

Davies, W. M. *The Gospel and the Land: Early Christianity and Jewish Territorial Doctrine.* Berkeley, 1974.

Dilke, O. A. W. "Illustrations from Roman Surveyors' Manuals." *Imago mundi* 21 (1967): 9–29.

———. "Maps in the Treatises of Roman Land Surveyors." *Geographical Journal* 127 (1961): 417–26.

Dillemann, Louis. *La Cosmographie du Ravennate.* Ed. Y. Janvier. Brussels, 1997.

Diller, A. "The Ancient Measurements of the Earth." *Isis* 40 (1949): 6–9.

Dorofeeva-Lichtman, Vera V. "Conception of Terrestrial Organization in the *Shan hai jing.*" *Bulletin de l'École Française de l'Extrême-Orient* 82 (1995): 57–110.

Eastwood, Bruce. *Astronomy and Optics from Pliny to Decartes: Texts, Diagrams, and Conceptual Structures.* London, 1989.

———. "Invention and Reform in Latin Planetary Astronomy of the Eleventh Century." In *Publications of the Journal of Medieval Latin,* ed. M. Herren. Turnhout, forthcoming.

———. "Latin Planetary Studies in the IXth and Xth Centuries." *Physis,* n.s., 32 (1995): 217–26.

Eder, Christine Elisabeth. "Die Schule des Klosters Tegernsee im frühen Mittelalter im Spiegel der Tegernsee Handschriften." *Studien und Mitteilungen zur Geschichte des Benediktiner-Ordens und seine Zweige* 83 (1972): 6–155. Reprinted as a monograph. Munich, 1972.

Edson, Evelyn. *Mapping Time and Space: How Medieval Mapmakers Viewed Their World.* London, 1997.

Edwards, Burton Van Name. "In Search of the Authentic Commentary on Genesis by Remigius of Auxerre." In *L'école carolingienne d'Auxerre de Murethach à Remi, 830–908,* ed. Dominique Iogna-Prat et al., 399–412. Paris, 1991.

———. "The Two Commentaries on Genesis Attributed to Remigius of Auxerre, with a Critical Edition of Stegmüller 7195." Ph.D. diss., University of Pennsylvania, 1992.

Englisch, Brigitte. *Die Artes liberales im frühen Mittelalter.* Stuttgart, 1994.

Epstein, Steven. "The Theory and Practice of the Just Wage." *Journal of Medieval History* 17 (1991): 53–69.

Esposito, Mario. "An Unpublished Astronomical Treatise by the Irish Monk Dicuil." *Proceedings of the Royal Irish Academy* 26 C (1907): 381–445.

Ewig, Eugen. "Die Civitas Ubiorum, die Francia Rinensis und das Land Ribuarien." *Rheinische Vierteljahrsblätter* 19 (1954): 1–29.

Fabrini, F. *Paolo Orosio, uno storico.* Rome, 1974.

Famulus Christi: Essays in Commemoration of the Thirteenth Centenary of the Birth of the Venerable Bede. Ed. Gerald Bonner. London, 1976.

Fernandez Caton, Jose Maria. *Las Etimologias en la tradicion manuscrita medieval.* Leon, 1966.

Folkerts, Menso. "The Importance of the Pseudo-Boethian *Geometria.*" In *Boethius and the Liberal Arts,* 187–209. Bern, 1981.

Fontaine, Jacques. "Cohérence et originalité de l'étymologie isidorienne." In *Homenaje a Eleuterio Elorduy, S.J.,* 113–44. Bilbao, 1978. Reprint in Jacques Fontaine, *Tradition et actualité chez Isidore de Seville,* chap. 10. London, 1988.

———. "Isidore de Seville et la mutation de l'encyclopédisme antique." In *La pensée encyclopédique au Moyen Age,* 519–38. Neuchâtel, 1966. Reprint in Jacques Fontaine, *Tradition et actualité chez Isidore de Seville,* chap. 4. London, 1988.

———. *Isidore of Seville et la culture classique dans l'Espagne wisigothique.* 2 vols. Paris, 1959.

———. "La culture carolingienne dans les abbayes normandes: L'exemple de Saint-Wandrille." In *Aspects du monachisme en Normandie, IVe–XVIIIe siècles: Actes du colloque scientifique de l' "Année des abbayes normandes,"* 31–54. Paris, 1962.

———. "La diffusion carolingienne du *De natura rerum* d'Isidore de Seville d'après les manuscrits conservés en Italie." *Studi medievali,* 3d ser., 7 (1966): 108–27.

Fuson, Robert H. *A Geography of Geography: Origins and Development of the Discipline.* Dubuque, Iowa, 1969.

Ganz, David. *Corbie in the Carolingian Renaissance.* Sigmaringen, 1990.

Gautier Dalché, Patrick. "De la glose à la contemplation: Place et fonction de la carte dans les manuscrits du haut Moyen Age." In *Testo e immagine nell'alto medioevo,* 2:693–764. Spoleto, 1994.

———. *La "Descriptio mappe mundi" de Hugues de Saint-Victor: Texte inédit avec introduction et commentaire.* Paris, 1988.

———. "La géographie descriptive à Saint-Germain d'Auxerre (milieu IXe–début Xe siècle)." In *Saint-Germain d'Auxerre: Intellectuels et artistes dans l'Europe carolingienne, IX–XI siècles,* 270–76. Auxerre, 1990.

———. "L'espace de l'histoire: Le rôle de la géographie dans les chroniques universelles." In *L'Historiographie médiévale en Europe: Actes du colloque, Paris, 29 mars–1er avril 1989,* ed. J.-Ph. Genet, 287–300. Paris, 1991.

———. "*Situs orbis terre vel regionum:* Un traité de géographie inédit du haut Moyen Age (Paris, B.N. latin 4841)." *Revue d'histoire des textes* 12–13 (1982–83): 149–79.

———. "Tradition et renouvellement dans la représentation d'espace géographique au IXe siècle." *Studi medievali,* 3d ser., 24 (1983): 121–65.

———. "Un problème d'histoire culturelle: Perception et représentation de l'espace au Moyen Age." *Médiévales: Langues, textes, histoire* 18 (1990): 5–15.

Gilson, Etienne. *History of Christian Philosophy in the Middle Ages.* New York, 1955.

Glauche, Günter. *Schullektüre im Mittelalter: Entstehung und Wandlungen des Lektürekanons bis 1200 nach den Quellen dargestellt.* Munich, 1970.

Goffart, Walter. *The Narrators of Barbarian History (A.D. 550–800): Jordanes, Gregory of Tours, Bede, and Paul the Deacon.* Princeton, 1988.

Gorman, Michael. "The Encyclopedic Commentary on Genesis Prepared for Charlemagne by Wigbod." *Recherches augustiniennes* 17 (1982): 173–201.

Gormley, Catherine M., Mary A. Rouse, and Richard H. Rouse. "The Medieval Circulation of the *De Chorographia* of Pomponius Mela." *Mediaeval Studies* 46 (1984): 266–320.

Graf, Arturo. *La leggenda del paradiso terrestre.* Turin, 1878.

Gregory, Derek. *Geographical Imaginations.* Oxford, 1994.

Grierson, Philip. "Abbot Fulco and the Date of the *Gesta abbatum Fontanellensium.*" *English Historical Review* 55 (1940): 275–84.

Grimm, Reinhold R. *Paradisus coelestis, Paradisus terrestris: Zur Auslegungsgeschichte des Paradieses im Abendland bis um 1200.* Munich, 1977.

Guenée, Bernard. *Histoire et culture historique dans l'Occident médiéval.* Paris, 1980.

Günther, Siegmund. *Geschichte der Erdkunde.* Leipzig and Vienna, 1904.

Gurevich, Aaron. *Categories of Medieval Culture.* Trans. G. Campbell. London and Boston, 1985.

Hanning, Robert W. *The Vision of History in Early Britain, from Gildas to Geoffrey of Monmouth.* New York and London, 1966.

Hay, Denys. *Europe: The Emergence of an Idea.* Edinburgh, 1957.

Herren, Michael W. "The Commentary on Martianus Attributed to John Scottus: Its Hiberno-Latin Background." In *Jean Scot écrivain: Acts du IVe colloque international, Montréal, 28 août–2 septembre 1983,* ed. G.-H. Allard, 265–86. Montreal and Paris, 1986.

———. "Wozu diente die Fälschung der Kosmographie des Aethicus?" In *Lateinische Kultur im VIII. Jahrhundert: Traube-Gedenkschrift,* ed. Albert Lehner and Walter Berschin, 145–59. St. Ottilien, 1989.

Hettner, Alfred. *Die Geographie: Ihre Geschichte, ihr Wesen und ihre Methoden.* Breslau, 1927.

Hexter, Ralph. *Ovid and Medieval Schooling: Studies in Medieval School Commentaries on Ovid's "Ars Amatoria," "Epistulae ex Ponto," and "Epistulae Heroidum."* Munich, 1986.

Heyse, Elisabeth. *Hrabanus Maurus' Enzyklopädie "De rerum naturis": Untersuchungen zu den Quellen und zur Methode der Kompilation.* Munich, 1969.

Higounet, Charles. "A propos de la perception de l'espace au Moyen Age." In *Media in Francia,* 257–68. Maulévrier, 1989.

Hillgarth, J. H. "Historiography in Visigothic Spain." In *La storiografia altomedievale,* 261–311. Settimane di Centro Italiano di studi sull'alto medioevo, 17. Spoleto, 1970.

Hillkowitz, Kurt. *Zur Kosmographie des Aethicus.* Vol. 1. Cologne, 1934. Vol. 2. Frankfurt am Main, 1973.

The History of Cartography. Vol. 1, *Cartography in Prehistoric, Ancient, and Medieval Europe and the Mediterranean.* Ed. J. B. Harley and David Woodward. Chicago, 1987.

Holt-Jensen, Arild. *Geography: Its History and Concepts.* Trans. Brian Fullerton. London, 1982.

Hunt, E. D. *Holy Land Pilgrimage in the Later Roman Empire, A.D. 312–460.* Oxford, 1982.

Hunter Blair, Peter. *Bede's Ecclesiastical History of the English Nation and Its Importance Today.* Jarrow Lecture, 1959. Newcastle-upon-Tyne, 1959.

Iogna-Prat, Dominique. "L'oeuvre d'Haymon d'Auxerre: État de la question." In *L'école carolingienne d'Auxerre de Murethach à Remi, 830–908,* ed. Dominique Iogna-Prat et al., 157–79. Paris, 1991.

Janvier, Yves. *La géographie d'Orose.* Paris, 1982.

Jean Scot Erigène et l'histoire de la philosophie: Colloque Internationale du Centre de la Recherches Scientifiques, Laon, 7–12 juillet 1975. Ed. René Roques. Paris, 1977.

Jeauneau, Edouard. "Heiric d'Auxerre disciple de Jean Scot." In *L'école carolingienne d'Auxerre de Murethach à Remi, 830–908,* ed. Dominique Iogna-Prat et al., 353–70. Paris, 1991.

Jones, Charles W. "Some Introductory Remarks on Bede's Commentary on Genesis." *Sacris Erudiri* 19 (1969–70): 115–98. Reprinted in Charles W. Jones, *Bede, the Schools, and the Computus,* ed. Wesley M. Stevens (Aldershot, 1994), chap. 4.

Keane, John. *The Evolution of Geography.* London, 1899.

Kelly, J. N. D. *Jerome: His Life, Writings, and Controversies.* New York, 1975.

Kendall, Calvin B. "Imitation and the Venerable Bede's *Historia Ecclesiastica.*" In *Saints, Scholars, and Heroes: Studies in Medieval Culture in Honor of Charles W. Jones,* ed. Margot H. King and Wesley M. Stevens, 1:161–90. Collegeville, 1979.

Kimble, George H. T. *Geography in the Middle Ages.* London, 1938.

Klinck, Roswitha. *Die lateinische Etymologie des Mittelalters.* Munich, 1970.

Klotz, Alfred. "Beiträge zur Analyse des geographischen Kapitels in Geschichtswerk des Orosius (I.2)." In *Charisteria A. Rzach,* 120–30. Reichenberg, 1930.

Körtum, Hans-Henning. *Richer von Saint-Remi: Studien zu einem Geschichtsschreiber des 10. Jahrhunderts.* Stuttgart, 1985.

Kretschmer, Konrad. *Die physische Erdkunde im christlichen Mittelalter: Versuch einer quellenmäßigen Darstellung ihrer historishen Entwicklung.* Vienna, 1889.
Kupfer, Marcia. "Medieval World Maps: Embedded Images, Interpretive Frames." *Word and Image* 10 (1994): 262–88.
Kuznetsova, Natalia. *Nauka v ee istorii* (Science in its history). Moscow, 1982.
Labowsky, Lotte. "A New Version of Scotus Eriugena's Commentary on Martianus Capella." *Mediaeval and Renaissance Studies* 1 (1941–43): 187–93.
Laistner, M. L. W. *Thought and Letters in Western Europe: A.D. 500 to 900.* London, 1931.
La storiografia altomedievale. Settimane di Centro Italiano di studi sull'alto medioevo, 17. Spoleto, 1970.
Latouche, Robert. "Un imitateur de Salluste au Xe siècle: L'historien Richer." *Annales de l'Université de Grenoble, nouvelle série, section Lettres-Droit* 6 (1929): 289–305.
L'école carolingienne d'Auxerre de Murethach à Remi, 830–908. Ed. Dominique Iogna-Prat et al. Paris, 1991.
Lee, A. D. *Information and Frontiers: Roman Foreign Relations in Late Antiquity.* Cambridge, 1993.
Le Goff, Jacques. *Medieval Civilization.* Trans. J. Barrow. New York, 1988.
Lelewel, Joachim. *La géographie du Moyen Âge.* 4 vols. Brussels, 1852.
Le métier d'historien au Moyen Age: Etudes sur l'historiographie médiévale. Ed. Bernard Guenée. Paris, 1977.
Leonardi, Claudio. "I codici di Marziano Capella." *Aevum* 33 (1959): 443–89; 34 (1960): 1–99, 411–524.
———. "Illustrazioni e glosse in un codice di Marziano Capella." *Archivio Paleografico Italiano,* n.s., 2–3, pt. 2 (1956–57): 39–60.
———. "Nota introduttiva per un'indagine sulla fortuna di Marziano Capella nel Medioevo." *Bulletino dell'Istituto storico italiano* 67 (1955): 265–88.
Lesne, Emile. *Les livres, scriptoria et bibliothèques du commencement du VIIIe siècle à la fin du XIIe siècle.* Lille, 1938.
Lindberg, David. *The Beginnings of Western Science.* Chicago, 1992.
———. "Science and the Early Christian Church." *Isis* 74 (1983): 509–30.
Lindgren, Uta. "Geographie in der Zeit der Karolinger." In *Karl der Grosse und sein Nachwirken: 1200 Jahre Kultur und Wissenschaft in Europa.* Vol. 1, *Wissen und Weltbild,* ed. P. Butzer, M. Kerner, and W. Oberschelp, 507–19. Turnhout, 1997.
Lipp, Frances. "The Carolingian Commentaries on Bede's *De natura rerum.*" Ph.D. diss., Yale University, 1961.
Livingstone, David. *The Geographical Tradition: Episodes in the History of a Contested Enterprise.* Oxford, 1992.
Löwe, Heinz. "Die 'Vacetae insolae' und die Entstehungszeit der Kosmographie des Aethicus Ister." *Deutsches Archiv für Erforschung des Mittelalters* 31 (1975): 1–16.
———. "Ein literarischer Widersacher des Bonifatius: Virgil von Salzburg und die Kosmographie des Aethicus Ister." *Abhandlungen der Akademie der Wis-*

senschaften und der Literatur in Mainz, Geistes- und sozialwissenschaftlichen Klasse 11 (1951): 903–83.

H. Löwe. "Salzburg als Zentrum literarischen Schaffens im 8. Jahrhundert." *Mitteilungen der Gesellschaft für Salzburger Landeskunde* 115 (1975): 99–143.

Löwenberg, Julius. *Geschichte der Geographie.* Berlin, 1840.

Lozovsky, Natalia. "Carolingian Geographical Tradition: Was It Geography?" *Early Medieval Europe* 5 (1996): 25–43.

———. "The Explanation of Geographical Material in the Commentary by Remigius of Auxerre." *Studi medievali,* 3d ser., 34 (1993): 563–72.

Lubac, Henri de. *Exégèse médiévale: Les quatres sens de l'Ecriture.* 4 vols. Paris, 1959–65.

Lüdecke, Friedrich. "De Marciani Capellae libro sexto." Ph.D. diss., Göttingen, 1862.

Lugge, Margret. *Gallia und Francia im Mittelalter.* Bonn, 1960.

Lutz, Cora. "Martianus Capella." In *Catalogus translationum et commentariorum: Medieval and Renaissance Latin Translations and Commentaries, Annotated Lists and Guides,* ed. Paul O. Kristeller and F. Edward Cranz, 2:367–81. Washington, 1971.

———. "Remigius' Ideas on the Origin of the Seven Liberal Arts." *Medievalia et humanistica* 10 (1956): 32–49.

Madec, G. "Observations sur le dossier augustinien du *Periphyseon.*" In *Eriugena: Studien zu seinen Quellen: Vorträge des III. Internationalen Eriugena-Colloquiums, Freiburg im Breisgau, 27.–30. August 1979,* ed. W. Beierwaltes, 75–84. Heidelberg, 1980.

Manitius, Max. *Geschichte der lateinischen Literatur des Mittelalters.* Vol. 1. Munich, 1911.

Marenbon, John. *From the Circle of Alcuin to the School of Auxerre.* Cambridge, 1981.

Mariétan, Joseph. *Problème de la classification des sciences d'Aristote à s. Thomas.* Paris, 1901.

Markus, R. A. "Marius Victorinus and Augustine." In *The Cambridge History of Later Greek and Early Medieval Philosophy,* ed. A. H. Armstrong, 329–419. Cambridge, 1970.

Marrou, Henri I. "Saint Augustin, Orose et l'augustinisme historique." In *La storiografia altomedievale,* 59–87. Settimane di Centro Italiano di studi sull'alto medioevo, 17. Spoleto, 1970.

Martianus Capella and the Seven Liberal Arts. 2 vols. Vol. 1, *The Quadrivium of Martianus Capella: Latin Traditions in the Mathematical Sciences, 50* B.C.-A.D. *1250,* by William Harris Stahl, with *a Study of the Allegory and the Verbal Disciplines,* by R. Johnson with E. L. Burge. Vol. 2, *The Marriage of Philology and Mercury,* translated by William Harris Stahl and Richard Johnson, with E. L. Burge. New York, 1977.

Martin, Geoffrey J., and Preston E. James. *All Possible Worlds.* 3d ed. New York, 1993.

Matter, Ann. "Exegesis and Christian Education: The Carolingian Model." In

Schools of Thought in the Christian Tradition, ed. Patrick Henry, 90–105. Philadelphia, 1984.

McClure, Judith. "Bede's Old Testament Kings." In *Ideal and Reality in Frankish and Anglo-Saxon Society,* ed. Patrick Wormald, 76–129. Oxford, 1986.

McCluskey, Stephen C. *Astronomies and Cultures in Early Medieval Europe.* Cambridge, 1998.

McKitterick, Rosamond. *The Carolingians and the Written Word.* Cambridge, 1989.

———. "Knowledge of Plato's *Timaeus* in the Ninth Century: The Implications of Valenciennes, Bibliothèque Municipale Ms 293." In *From Athens to Chartres: Neoplatonism and Mediaeval Thought,* ed. H. J. Westra, 85–95. Leiden, 1992. Reprinted in Rosamond McKitterick, *Books, Scribes, and Learning in the Frankish Kingdoms, 6th–9th Centuries* (Aldershot, 1994), chap. 10.

Meier, Christel. "Das Problem der Qualitätenallegorese." *Frühmittelalterliche Studien* 8 (1974): 385–435.

Melón, Amando. "La etapa isidoriana en la geografía medieval." *Arbor* 28 (1954): 456–67.

Milham, Mary Ella. "A Handlist of the Manuscripts of C. Julius Solinus." *Scriptorium* 37 (1983): 126–29.

The Mind of Eriugena: Papers of a Colloquium, Dublin, 14–18 July 1970. Ed. John J. O'Meara and Ludwig Bieler. Dublin, 1973.

Mommsen, Theodor E.. "Orosius and Augustine." In *Medieval and Renaissance Studies,* ed. Eugene F. Rice, 325–48. Ithaca, 1959.

Müllenhoff, Karl. *Über die Weltkarte und Chorographie des Kaisers Augustus.* Kiel, 1856.

Munk Olsen, B. *L'étude des auteurs classiques latins aux XIe et XIIe siècles.* Vol. 2. Paris, 1985.

Nature and Science: Essays in the History of Geographical Knowledge. N.p., 1992.

Nicolet, Claude, and Patrick Gautier Dalché. "Les 'quatre sages' de Jules César et la 'mesure du monde' selon Julius Honorius: Réalité antique et tradition médiévale." *Journal des savants* (1986): 157–218.

O'Donnell, James J. "The Aims of Jordanes." *Historia* 31 (1982): 223–40.

———. *Cassiodorus.* Berkeley, Los Angeles, and London, 1979.

Ohly, Friedrich. "Vom geistigen Sinn des Wortes im Mittelalter." In *Schriften zur mittelalterlichen Bedeutungsforschung,* 1–31. Darmstadt, 1977.

O'Loughlin, Thomas. "The Exegetical Purpose of Adomnan's *De locis sanctis.*" *Cambridge Medieval Celtic Studies* 24 (winter 1992): 37–53.

Optima hereditas: Sapienza giuridica romana e conscenza dell'ecumene. Ed. Germaine Aujac et al. Milan, 1992.

Patch, Howard Rollin. *The Other World according to Descriptions in Medieval Literature.* Cambridge, 1950.

Peschel, Otto. *Geschichte der Erdkunde.* 2d ed. Munich, 1877.

Philipp, Hans. *Die historisch-geographischen Quellen in den Etymologiae des Isidorus von Sevilla.* Berlin, 1912.

Préaux, Jean. "Deux manuscrits gantois de Martianus Capella." *Scriptorium* 13 (1959): 15–21.

———. "Le commentaire de Martin de Laon sur l'oeuvre de Martianus Capella." *Latomus* 12 (1953): 437–59.

———. "Les manuscrits principaux du *De nuptiis Philologiae et Mercurii* de Martianus Capella." In *Lettres latins du Moyen Age et de la Renaissance*, 76–128. Brussels, 1978.

Quadri, Riccardo. "Aimone di Auxerre alla luce dei 'Collectanea' di Heiric di Auxerre." *Italia medioevale e umanistica* 6 (1963): 1–48.

Quain, Edwin A. *The Medieval Accessus ad Auctores*. New York, 1986.

Rabinovitch, Vadim. *Alkhimiya kak fenomen srednevekovoy kultury* (Alchemy as a phenomenon of medieval culture). Moscow, 1979.

Ray, Roger D. "Bede, the Exegete, as Historian." In *Famulus Christi: Essays in Commemoration of the Thirteenth Centenary of the Birth of the Venerable Bede*, ed. Gerald Bonner, 125–40. London, 1976.

Reynolds, Suzanne. *Medieval Reading: Grammar, Rhetoric, and the Classical Text*. Cambridge, 1996.

Richard, J. "Les relations de pèlerinage et les motivations de leurs auteurs." In *Wallfahrt kennt keine Grenzen: Ausstellung im Bayerischen Nationalmuseum, München, 28. Juni bis 7. Oktober 1984*, ed. Thomas Raff, 143–54. Munich and Zurich, 1984.

———. "Voyages réels et voyages imaginaires: Instruments de la connaissance géographique au Moyen Age." In *Culture et travail intellectuel dans l'Occident médiéval*, 211–20. Paris, 1981.

Riché, Pierre. *Education and Culture in the Barbarian West, Sixth through Eighth Centuries*. Trans. John J. Contreni. Columbia, S.C., 1976.

———. *Les écoles et l'enseignement dans l'Occident chrétien de la fin du Ve siècle au milieu du XIe siècle*. Paris, 1979.

Ringbom, L. I. *Paradisus terrestris: Myt, Bild och Verklighet*. Helsinki, 1958.

Ritschl, Friedrich. *De M. Terentii Varronis Disciplinarum libris commentarius*. In Friedrich Ritschl, *Kleine philologische Schriften (Opuscula philologica)*, 3:352–40. Leipzig, 1877. Reprint Hildesheim, 1978.

Romm, James S. *The Edges of the Earth in Ancient Thought: Geography, Exploration, and Fiction*. Princeton, 1992.

Rück, Karl. "Die Naturalis Historia des Plinius im Mittelalter: Exzerpte aus der Naturalis Historia auf den Bibliotheken zu Lucca, Paris und Leiden." *Sitzungsberichte der Königlich Bayerischen Akademie der Wissenschaften zu München, Philosophisch-historische Abteilung* 1 (1898): 203–318.

Schipper, William. "Rabanus Maurus: *De rerum naturis*: A Provisional Check List of Manuscripts." *Manuscripta* 33 (1989): 109–18.

Schmale, Franz-Josef. *Funktion und Formen mittelalterlicher Geschichtsschreibung: Eine Einführung, mit einem Beitrag von Hans-Werner Goetz*. Darmstadt, 1985.

Schmithüsen, Josef. *Geschichte der geographischen Wissenschaft von den ersten Anfängen bis zum Ende des 18. Jahrhunderts*. Mannheim, Vienna, and Zurich, 1970.

Schneider, A. *Die Erkenntnislehre des Johannes Eriugena im Rahmen ihrer metaph-*

ysischen und anthropologischen Voraussetzungen nach den Quellen dargestellt. 2 vols. Berlin and Leipzig, 1921–23.

Schnetz, Joseph. "Jordanis beim Geographen von Ravenna." *Philologus* 81 (1925–26): 86–100.

———. *Untersuchungen über die Quellen der Kosmographie des anonymen Geographen von Ravenna.* Munich, 1942.

———. *Untersuchungen zum Geographen von Ravenna.* Munich, 1919.

———. "Zur Beschreibung des Alamannenlandes beim Geographen von Ravenna." *Zeitschrift für Geschichte des Oberrheins* 75, n.s., 36 (1921): 337–41.

Schools of Thought in the Christian Tradition. Ed. Patrick Henry. Philadelphia, 1984.

Schrimpf, Gangolf. "Die Sinnmitte von Periphyseon." In *Jean Scot Erigène et l'histoire de la philosophie: Colloque Internationale du Centre de la Recherches Scientifiques, Laon, 7–12 juillet 1975,* ed. René Roques, 289–305. Paris, 1977.

———. *Das Werk des Johannes Scottus Eriugena im Rahmen des Wissenschaftsverständnisses seiner Zeit: Eine Hinführung zu 'Periphyseon.'* Munster, 1982.

———. "Zur Frage der Authentizität unserer Texte von Johannes Scottus' *Annotationes in Martianum.*" In *The Mind of Eriugena: Papers of a Colloquium, Dublin, 14–18 July 1970,* ed. John J. O'Meara and Ludwig Bieler, 49–58. Dublin, 1973.

Schwartz, Alexander. "Glossen als Texte." *Beiträge zur Geschichte der deutschen Sprache und Literatur* 99 (1977): 25–36.

Science in the Middle Ages. Ed. David Lindberg. Chicago, 1978.

Science in Western and Eastern Civilization in Carolingian Times. Ed. Paul. L. Butzer and Dietrich Lohrmann. Basel, Boston, and Berlin, 1993.

The Seven Liberal Arts in the Middle Ages. Ed. David L. Wagner. Bloomington, 1983.

Shanzer, Danuta. *A Philosophical and Literary Commentary on Martianus Capella's "De nuptiis Philologiae et Mercurii" Book 1.* Berkeley, Los Angeles, and London, 1986.

Simon, Manfred. "Zur Abhängigkeit spätrömischer Enzyklopädien der Artes Liberales von Varros Disciplinarum libri." *Philologus* 110 (1966): 88–101.

Smalley, Beryl. "The Bible in the Medieval Schools." In *The Cambridge History of the Bible,* vol. 2, *The West from the Fathers to the Reformation,* ed. G. W. H. Lampe, 197–220. Cambridge, 1969.

———. *The Study of the Bible in the Middle Ages.* Oxford, 1941.

Smyth, Marina. *Understanding the Universe in Seventh-Century Ireland.* Woodbridge, 1996.

Sot, Michel. "Organization de l'espace et historiographie épiscopale dans quelques cités de la Gaule carolignenne." In *Le métier d'historien au Moyen Age: Etudes sur l'historiographie médiévale,* ed. Bernard Guenée, 31–43. Paris, 1977.

Spicq, C. *Esquisse d'une histoire de l'exégèse latine au Moyen Age.* Paris, 1944.

Spitzer, Leo. "The Epic Style of the Pilgrim Aetheria." In *Romanische Literaturstudien 1936–1956,* 871–912. Tübingen, 1959.

Staab, Franz. "Ostrogothic Geographers at the Court of Theodoric the Great: A

Study of Some Sources of the Anonymous Cosmographer of Ravenna." *Viator* 7 (1976): 27–58.

Stahl, William H. "Dominant Traditions in Early Medieval Latin Science." *Isis* 50 (1959): 95–124.

———. "The Quadrivium of Martianus Capella: Latin Traditions in the Mathematical Sciences, 50 B.C.–A.D. 1250." In *Martianus Capella and the Seven Liberal Arts.* New York, 1977.

———. *Roman Science: Origins, Development, and Influence to the Later Middle Ages.* Madison, 1962.

Stevens, Wesley M. *Cycles of Time and Scientific Learning in Medieval Europe.* Aldershot, 1995.

Sumption, Jonathan. *Pilgrimage: An Image of Medieval Religion.* Totowa, 1975.

Svennung, Josef. *Jordanes und Scandia: Kritischexegetische Studien.* Stockholm, 1967.

Swain, J. W. "The Theory of the Four Monarchies." *Journal of Classical Philology* 35 (1940): 1–21.

Szerwiniack, Olivier. "Un commentaire hiberno-latin des deux premiers livres d'Orose, *Histoire contre les pagans.*" *Archivium latinitatis Medii Aevi (Bulletin Du Cange)* 51 (1992–93): 5–137.

Taylor, Joan F. *Christians and the Holy Places: The Myth of Jewish-Christian Origins.* New York, 1993.

Text and Territory: Geographical Imagination and the European Middle Ages. Ed. Sylvia Tomasch and Sealy Gilles. Philadelphia, 1998.

Texts and Transmissions. A Survey of the Latin Classics. Ed. L. D. Reynolds. Oxford, 1983.

Thomson, J. Oliver. *History of Ancient Geography.* New York, 1965.

Travel and Travellers in the Middle Ages. Ed. A. P. Newton. New York, 1962.

Uhden, Richard. "Die Weltkarte des Isidorus von Sevilla." *Mnemosyne* 3, no. 1 (1935): 1–28.

Ullman, B. L. "Geometry in the Medieval Quadrivium." In *Studi di bibliografia e di storia in onore di Tammaro de Marinis,* 4:263–85. Verona, 1964.

Vaugondy, Robert de. *Essai sur l'histoire de la géographie.* Paris, 1755.

Vermeer, G. F. M. *Observations sur le vocabulaire du pèlerinage chez Egérie et chez Antonin de Plaisance.* Nijmegen, 1965.

Volkov, Alexei. "Science and Daoism: An Introduction." *Taiwanese Journal for Philosophy and History of Science* 8 (1996–97): 1–58.

Wagner, Norbert. *Getica: Untersuchungen zum Leben des Jordanes und zur frühen Geschichte der Goten.* Berlin, 1967.

Wallace-Hadrill, J. M. *Bede's Ecclesiastical History of the English People: A Historical Commentary.* Oxford, 1988.

Wallis, Faith. "The Church, the World, and the Time: Prolegomena to a History of the Medieval *Computus.*" In *Normes et pouvoirs à la fin du Moyen Age,* ed. M.-C. Désprez-Masson, 15–29. Montreal, 1990.

———. "Images of Order in the Medieval *Computus.*" In *Ideas of order in the Middle Ages.* Ed. Warren Ginsberg, 45–68. Binghamton, 1990.

———. "Ms Oxford St. John's College 17: A Mediaeval Manuscript in Its Context." Ph.D. diss., University of Toronto, 1985.

Ward, Benedicta. *Miracles and the Medieval Mind: Theory, Record, and Event, 1000–1215.* Philadelphia, 1987.

Weibull, Lauritz. "Scandza und ihre Völker in der Darstellung des Jordanes." *Arkiv för Nordisk Filologi* 41 (1925): 213–46.

Weisheipl, James A. "Classification of the Sciences in Medieval Thought." *Mediaeval Studies* 27 (1965): 63–64.

White, Alison M. "Glosses Composed before the Twelfth Century in Manuscripts of Macrobius' Commentary on Cicero's Somnium Scipionis." Ph.D. diss., Oxford University, 1981.

Wilkinson, John. *Jerusalem Pilgrims before the Crusades.* Warminster, 1977.

Willis, James A. "Martianus Capella and His Early Commentators." Ph.D. diss., University of London, 1952.

———. "Martianus Capella und die mittelalterliche Schulbildung." *Das Altertum* 19 (1973): 164–74.

Witzel, Hans-Joachim. *Der geographische Exkurs in den lateinischen Geschichtsquellen des Mittelalters.* Frankfurt, 1952.

Wolfram, Herwig. *History of the Goths.* Trans. Thomas J. Dunlap. Berkeley and London, 1988.

Wright, John K. *The Geographical Lore of the Time of the Crusades: A Study in the History of Medieval Science and Tradition in Western Europe.* New York, 1925.

———. *Human Nature in Geography.* Cambridge, 1966.

Zangemeister, Karl. "Die Chorographie des Orosius." In *Commentationes philologae in honorem Th. Mommseni,* 715–38. Berlin, 1877.

Zeiler, Michaela. "'Quicumque aut quilibet sapiens Aethicum aut Mantuanum legerit' Muß der Name des Verfassers der Kosmographie wirklich 'in geheimnisvolles Dunkel gehüllt bleiben'?" *Wiener Studien* 104 (1991): 183–207.

Zimmerman, Michel. "Le monde d'un catalan au Xe siècle: Analyse d'une compilation isidorienne." In *Le métier d'historien au Moyen Age: Etudes sur l'historiographie médiévale,* ed. Bernard Guenée, 45–79. Paris, 1977.

Zumthor, Paul. *La mésure du monde: Representation de l'espace au Moyen Age.* Paris, 1993.

General Index

Ablavius, 81–82
Abraham, 40–46, 49–50, 59, 134, 143
Adamnan, 48
Aethicus Ister, 28, 31–33, 150–52
Africa, 40, 41, 56, 71, 74, 75, 80, 83n. 73, 95, 105, 106, 108, 110, 134, 136
Agrippa, 72
Alcuin, 43
Alexander the Great, 60, 76, 146
Alexandria, 42
Amazons, 76, 85n. 83
Ambrose of Milan, 40–41
Ammianus Marcellinus, 8
Anaxagoras, 121
Angelomus of Luxeuil, 43
Anonymus Cantabriensis, 117–18, 127, 132, 133, 134, 136
Anonymus Leidensis, 78n. 49, 109–10, 115, 145, 147, 150, 152, 154
Anonymus Ravennatis, 30–31, 86, 145–46, 147
  on Paradise, 59–60
  on recent political divisions, 140
  on Scandza, 82
Archulf, 48
Armenia, 54, 56, 60, 146
Asia, 53, 54, 55, 71, 74, 76, 80, 83n. 73, 95, 105, 108, 112, 134
Assyria, 105, 111
Astronomy, 5n. 12, 12n. 15, 15, 18n. 42, 36n. 3

astronomical excerpts, 129
Augustine, Saint, 4, 6, 13n. 20, 16, 20, 21, 22, 69–70, 75n. 31, 77
  epistemology of, 141–43, 144
  explanation of Gen. 13:14, 42–43, 44, 45, 46
  justification of geographical studies, 10–14
  on Paradise, 51–53, 54, 55, 58, 59
Augustus, 72, 75, 97n. 139
Auxerre, 114, 115, 117, 119, 130

Bede, 21, 69
  on biblical topography, 48–50
  commentary on, 129–30, 138
  explanation of Gen. 13:14, 44–45
  geographical descriptions in *Historia Ecclesiastica*, 86–94, 100, 101, 139
  on Paradise, 55–56, 57, 58–59
  praise of Britain, 65–66, 89–90
Benedict, Saint, 37, 38
Bernard (ninth-century pilgrim), 47
Besançon, 99, 100
Boethius, 13n. 21, 14, 15, 18n. 42, 19, 23, 26, 28
Britain, 61, 100, 140, 148
  in Bede, 65–66, 87–90, 91, 94, 101
  in Gildas, 40, 64–65, 87
  in Jordanes, 80–81, 83–85
Byzantium. *See* Constantinople, called Byzantium

179

Caesar, Julius, 12, 88n. 98, 89, 96, 97, 136, 141, 149
Canaan, 45, 50, 105
Canterbury school commentaries, 61–62
Carthage, 42, 72–73, 75, 77
Cassiodorus, 6, 9, 20, 23, 26, 29n. 90, 79, 121n. 68
  and geographical studies at Vivarium, 7n. 2, 14–19
Censorinus, 18–19
Charlemagne, 44, 141, 148
Christian of Stavelot (Christian Druthmar), 37
Cicero, 8, 14
Computus (calendrical computation), 3n. 9, 5n. 12, 14, 29, 33
Constantinople, 17, 78
  called Byzantium, 73, 136–37, 141

Danube, 62
Dicuil, 28, 29, 78n. 49, 117
  and contemporary realities, 140, 141
  and eyewitness reports, 147–49
  and written authority, 148–49, 152
Dionysius Peiegetes, 17

Earth
  shape of, 7, 21, 22, 120–21, 123
  size of, 7, 22, 27, 120, 123, 127–32
  measurement of, 18, 19, 23, 25, 26, 27, 28, 29, 120, 123–24, 127–32
Egeria, 46–47
Egypt, 41, 49, 50, 56, 57, 58, 59, 60, 110, 129
Eratosthenes, 27, 123–30
Etymology, 48–49, 92, 95, 152, 154. *See also* Hrabanus Maurus, methods of; Isidore of Seville, methods of
Euphrates, 41, 51, 52, 53–54, 55, 56, 60
Europe, 61, 62, 71, 74, 79, 80, 83n. 73, 88, 95, 105, 106, 108, 134, 140
Eusebius, 11, 16n. 32, 48

Fidelis (ninth-century pilgrim), 148–49

Flodoard, 96n. 137, 98, 99–100
Freculf of Lisieux, 86, 147

Gaul, 61, 64, 88, 94, 95, 97, 98, 99, 101, 136, 154
  tripartite division of, 94, 96, 136, 141, 154
Gauls, 94, 96–97
Geometry, 15, 18, 19, 23–28, 33, 120, 128n. 85, 132
Geon, 51, 55, 59, 60
  and Ganges, 133
  in Gaul, 61, 64
  and Nile, 52, 53, 54n. 74, 56, 60
Gildas, 39–40, 64–65, 87, 88, 90n. 105
Gothia, 97, 105, 107
Goths, 65, 69, 72n. 16, 79, 82–83, 85–86, 107, 141
Grammar, 24, 29, 33, 111, 158
  explanation of, 120, 132, 137
Gregory of Tours, 25, 29n. 90, 68n. 1, 87n. 92, 148n. 35
Gregory the Great, 37, 38, 57n. 88, 58, 90–91

Haimo of Auxerre, 38, 44
Ham, 105, 106
Holy Land, 40–50, 61–62, 66. *See also* Palestine
Horace, 45, 99–100
Hrabanus Maurus, 21–22, 27, 29n. 90, 39, 49n. 56, 56–58
  methods of, 110–12, 154

Iaphet, 105, 106, 151
Isidore of Seville, 22, 29n. 90, 32, 56, 57, 58, 59, 78n. 49, 95, 133, 134, 137, 138, 140, 141, 145, 149, 150, 154
  geographical material in works of, 20–21, 26–27, 103–8
  influence of, 102, 107–13, 152
  methods of, 104–7, 132, 154
  on Paradise, 53–55, 62
  praise of Spain, 65, 68n. 1
Isle of Wight, 89, 91

Jerome, Saint, 16n. 32, 28, 31, 32, 33n. 107, 53, 96, 150
  on biblical topography, 48–50
  explanation of Gen. 13:14, 41, 43, 44
Jerusalem, 17, 45, 46n. 39, 47, 61, 62, 73, 112, 148
John Scottus Eriugena, 13n. 20, 22, 117
  and commentary on Martianus Capella, 113, 114, 121, 123, 129, 133, 134
  on earth measurement, 27, 124, 127, 130–31
  epistemology of, 143–44
Jordanes, 69, 89, 95
  geographical descriptions in *Getica,* 78–86, 100, 101, 139, 147, 148
Joseph and pyramids, 148
Josephus Flavius, 48, 52n. 67, 82
Julius Honorius, 9n. 7, 17, 28, 80n. 58

Land surveying, 18, 19, 27, 31
Laon, 117, 130
Libya, 105
Lucan, 133

Macrobius, 128n. 85, 129
Magog, 82n. 69, 105, 107
Maps, 4, 14, 17, 19n. 47, 33n. 107, 38, 39, 40, 55, 66, 71, 72, 80n. 58, 134–36
Marcellinus Comes, 16, 17
Martianus Capella, 5, 9, 29n. 90, 147, 150
  commentaries on, 113–38, 140, 141 (*see also* Anonymus Cantabriensis; John Scottus Eriugena, and commentary on Martianus Capella; Martin of Laon; Remigius of Auxerre, and commentaries on Martianus Capella)
  and geography in quadrivium, 23–28
Martin of Laon, 113, 114, 120, 124n. 75, 131, 132, 133, 134, 136n. 115
Mediterranean Sea, 106, 154

Mela, Pomponius, 7, 8, 10, 80–81, 82, 84n. 76, 116, 150, 154
Meroe, 130
Mesopotamia, 112

Nile, 52, 53, 54n. 74, 55, 56, 57, 58, 60, 64, 133, 134, 140, 148, 149
Noah, 31, 35n. 2, 49n. 56, 106, 107, 133

Orosius, 15, 39, 56, 80, 88, 89, 94, 95, 100, 107, 139, 150, 151, 153
  geographical descriptions in *Historiarum aduersum paganos,* 69–78, 83, 86, 87
  school texts based on, 107–8, 109, 117, 138

Palestine, 41–46, 49, 50, 66, 112. *See also* Holy Land
Paradise, 32, 35n. 2, 40, 50–59, 89, 94, 101, 107, 112, 133, 146
  allusions to, 63–66
  location of, 59–63
Philosophy, 13, 15, 20, 21, 22, 26, 141–45
Phison, 51, 59, 60
  and Ganges, 52, 53, 55, 56
  and Rhône and Danube, 62
Physica, 6, 20–22, 29, 33
Physiologia, 22
Pliny the Elder, 7, 8, 21, 24, 56, 80n. 54, 88, 107, 129–30, 136, 137, 138, 148, 149
Praise of places, 63, 65, 68n. 1, 89, 96
Pseudo-Aethicus, 28, 60–61, 64, 78n. 49
Ptolemy, 3n. 8, 7n. 2, 18, 81, 84, 127

Quadrivium, 6, 15, 19n. 47, 20, 23–28, 33, 132, 137

Reims, 28, 94, 107, 114, 117, 118, 119

Remigius of Auxerre, 44n. 31, 49n. 56, 56
  and commentaries on Martianus Capella, 113, 115n, 120n. 65, 121, 123, 124n. 75, 129, 133, 134
  and explanation of Gen. 13:14, 43
  on Paradise, 58–59
Rhetoric, 13–14, 29, 33, 137
  education in, 25
  tradition of, 71, 85, 99
Rhône, 62, 97
Richer, 69, 136, 139, 141
  geographical descriptions in *Historiae*, 94–100, 101
Roman Empire, 70, 74, 75, 76
Romans, 44, 89
Rome, 73, 74–75, 77

Sallust, 54
Scandza, 79–80, 81–85, 86, 100, 101, 147, 148
Scythia, 76, 82, 83, 85, 105
Scythians, 50, 82
Shem, 105, 106
Solinus, Julius, 7, 8, 24, 88, 107, 136, 137, 138, 148, 150
Strabo, 84
Sturmi, 153
Sundial, 27, 31, 123–29
  gnomon of, 123–24, 127–28, 129
Syene, 130

Tacitus, 12, 81, 84
Thule, 148, 149
Tigris, 51, 52, 54, 55, 56, 60, 61, 73n. 23, 146

Varro, 18, 24, 121n. 68

Wigbod and explanation of Gen. 13:14, 44